TH

Dr Anil Aggrawal is a known author of science books. He writes voraciously on science. His articles appear constantly in various national science magazines.

Dr Aggrawal teaches at the Maulana Azad Medical College, New Delhi.

THE RUPA BOOK OF BIOLOGY QUIZ

ANIL AGGRAWAL

RUPA

Published by
Rupa Publications India Pvt. Ltd 2004
7/16, Ansari Road, Daryaganj
New Delhi 110002

Sales centres:
Allahabad Bengaluru Chennai
Hyderabad Jaipur Kathmandu
Kolkata Mumbai

Copyright © Anil Aggrawal 2004

All rights reserved.
No part of this publication may be reproduced, transmitted,
or stored in a retrieval system, in any form or by any means, electronic,
mechanical, photocopying, recording or otherwise, without the prior
permission of the publisher.

ISBN: 978-81-716-7320-9

Sixth impression 2018

10 9 8 7 6

The moral right of the author has been asserted.

This book is sold subject to the condition that it shall not, by way
of trade or otherwise, be lent, resold, hired out, or otherwise circulated,
without the publisher's prior consent, in any form of binding or cover
other than that in which it is published.

*This book is respectfully dedicated to
Ms. Manju Gupta of the NBT
who encourages me to write more books*

Acknowledgements

I wish to thank my friends Mr. Y.S. Gill, and Pradip Bhattacharya of the *United Newspaper Network*, who helped me in various ways during the preparation of this book. My thanks are also due to Dr. G.P. Phondke, editor-in-chief of *Science Reporter*, who encourages me to write on science. During the preparation of this book, I had stimulating discussions with several knowledgeable people and friends, and the essence of those discussions has found its way in this book. Main among these friends are Dr. R.K. Sharma, Assistant Professor at the All India Institute of Medical Sciences, Dr. S.K. Sharma, Senior Medical Officer at the General Hospital, Gurgaon, Dr. Satbir Singh, Associate Professor of Radiology at the GB Pant Hospital, New Delhi, Dr. Basant Lal, Associate Professor of Forensic Medicine and Dr. S.K. Gupta, Associate Professor of Biochemistry at the MAMC. Several friends in the publishing industry revealed to me some finer nuances of book- writing. Whenever I had any problem, I asked for their help and they were most keen to help me. Main among them are Mr. Sudhir Bansal of Applied Media, Mr. Ravindra Saxena of Butterworths, Mr. Rajendra Sinha, Ms. Mohini Verma and Mr. C.B. Sharma of Oxford University Press, Mr. Pallava Bagla and Kollegala Sharma of *Science Reporter*, and Mrs. Deeksha Bisht of Vigyan Pragati. Several friends and relatives backed me up from foreign lands. Main among these are my sister Rewa Vasudeva and niece Kavita from Scotland, my brother Sunil from Italy, my brother-in-law Vicky from USA and my brothers-in-law Sanjiv and Sandeep and their wives Anita and Ritu from Australia. Prof. S.P. Gupta, formerly of Dyal Singh College, and Suryaprabh from Pune were major sources of inspiration for me. Other friends, who encouraged me, were Mr. Akhil Jain of Opal Advertising and Marketing, Mr. Hansraj Sharma of Dyal Singh College and Dr. P.K. Mukherjee of Deshbandhu College.

Several computer whiz-kids helped me sort out problems with my computer and word-processor. Main among them are Bhrigu Celly, Arvind Lamba, Vaneet Pasricha, Raja Sahi and my cousin Nalin Aggarwal. I just do not know how I would have handled my computer without them.

Last, but not the least, I am thankful, to my wife Aparna and my ten year old son Tarun, both of whom suffered patiently, while I was busy writing this book.

June 1995   Anil Aggrawal

Contents

1.	Story of Biology	1
2.	The Sciences	4
3.	The Cell	8
4.	Chromosomes and Genes	16
5.	The Proteins	21
6.	Vitamins	24
7.	The Origin of Life	26
8.	Micro-organisms	27
9.	Heredity and Evolution	32
10.	Diversity of Animal Life	37
11.	Diversity of Plant Life	65
12.	Great Experiments	82
13.	Great Biologists	86
14.	Biology in Everyday Life	94
15.	Great Books	103
16.	Living Beings and their Environment	105
17.	Curious Facts	106
18.	The Human Body	112
19.	Miscellaneous	116
20.	One-liners	124
21.	Photoquiz	125
22.	Picture Quiz	128
	Answers	133

1
STORY OF BIOLOGY

Ancient

1. What is the theory of spontaneous generation?
   (a) Biblical story of the creation of Adam (b) Origin of animals from the navel of God (c) Origin of life from non-life (d) The theory that man is generating his numbers in geometrical progression

2. The beginnings of biology were laid down when our ancestors curiously looked at the entrails of sacrificed animals. However, they used the knowledge for divination. Divination by looking at the liver of sacrificed animals was known as
   (a) hepatoscopy (b) gerontoscopy (c) sacrificiology (d) arteriology

3. Who gave the name artery to the blood vessels which carry oxygenated blood?
   (a) Hippocrates (b) Praxagoras of Cos (c) Herophilus (d) Erasistratus

4. Who was the first scientist to have conducted dissections of the human body?
   (a) Hippocrates (b) Plato (c) Aristotle (d) Alcmaeon

5. The early Chinese are supposed to have made lasting contributions in the fields of medicine and general biology. Which of the following contributions were made by the Chinese?
   (a) Ashes of sponges used to relieve goiter (b) Opium used to relieve pain (c) Ephedrine used for treating asthma (d) All of the above

6. He was one of the most renowned pupils of Plato. Plato called him 'the intelligence of the school'. Whom was he referring to?
   (a) Theophrastus (b) Pytheas (c) Aristotle (d) Strato

7. Identify this person.

8. Which scientist is reputed to have invented the term "biology"?
   (a) Aristotle (b) Hippocrates (c) Anton Van Leeuwenhock (d) Jean-Baptiste de Lamarck

9. Who was the first person to discover the existence of male and female reproductive cells?
   (a) Hippocrates (b) Aristotle (c) Hecataeus (d) Xenophanes

Medieval

10. Throughout the Middle Ages, the Church's disapproval of dissection virtually put a stop to anatomical studies. Who, in A.D. 1316, wrote the first book devoted to anatomy, thereby getting the coveted title of the 'restorer of anatomy'?
    (a) Mondino de' Luzzi (b) John Arderne (c) Guilemo Varignana (d) Bernard Gordon

11. Till mid-eighteenth century, it was widely believed that a living creature was preformed and perfect in every miniature detail in the egg or sperm. What were these little creatures called?
    (a) Spermacula (b) Ovo hominis (c) Homunculi (d) *Animaliens immaturiae*

12. Which biologist is best known for putting forth the 'biogenetic law', also known as the 'theory of recapitulation'?
    (a) John Kendrew (b) Ernst Haeckel (c) Konrad Von Gesner (d) Walther Flemming.

13. This famous scientist started a movement for free discussion which led to the formal organization of academies, notably the Royal Society of London. Who was this scientist?
    (a) Roger Bacon (b) Joseph Lister (c) Louis Pasteur (d) Francis Bacon

2
THE SCIENCES

14. Taxonomy is the study of
    (a) animal shapes (b) animal behaviour (c) plant growth (d) classification of organisms

15. The branch of biology that deals with the form and structure of organisms without regard to function is known as
    (a) morphology (b) morphogony (c) physiology (d) anatomy

16. The classification of organisms by their chemical constituents is known as
    (a) chemotaxis (b) chemotaxonomy (c) taxonomic anatomy (d) chemical classification

17. The science of the structure of animals and plants based upon dissection and microscopic observation is known as
    (a) physiology (b) biochemistry (c) anatomy (d) biophysics

18. The study of insects is known as
    (a) microzoology (b) pisciology (c) seriology (d) entomology

19. The branch of biology that studies organisms in terms of their relationships with other organisms vis-a-vis their local environment is known as
    (a) palaeontology (b) ethology (c) ecology (d) biochemistry

20. The science of the forms of life existing in prehistoric times as represented by fossil animals and plants is known as

(a) palaeontology (b) archaic biology (c) extinctology (d) pharmacology

21. The study of the relationship of living things to environment and each other in prehistoric times is known as
    (a) archaic ecology (b) palaeoecology (c) palaeobotany (d) palaeozoology

22. The dating of fossil animals and plants, by counting the ridges on fossil shells and corals, is known as
    (a) corallology (b) sporology (c) palaeochronology (d) dendrochronology

23. The study of annual rings in trees to determine dates and environmental conditions in the past is known as
    (a) chronology (b) dendrochronology (c) dendroclimatology (d) dendrology

24. The study of rainfall and other aspects of past climatic conditions by analysing the annual growth rings of trees is known as
    (a) dendroclimatology (b) dendrology (c) metereology (d) phocology

25. The branch of botany which deals only with the study of trees is known as
    (a) biochemistry (b) pathology (c) dendrology (d) pharmacology

26. The branch of palaeontology dealing with fossil plants is known as
    (a) palaeobotany (b) fossilology (c) pharmacology (d) pathology

27. The branch of palaeontology that deals with the biochemical constituents of fossil organisms is known as
    (a) palaeobiology (b) palaeobiochemistry (c) palaeology (d) palaeopharmacology

28. The study of patterns of distribution of animals and plants in different parts of the earth is known as
    (a) geology (b) biogeology (c) biogeography (d) biozoogeography

29. The study of plant spores and pollen, especially in fossil form, is known as
    (a) palynology (b) phocology (c) topology (d) palaeopollenology

30. The branch of biology which deals with the search for and study of extraterrestrial living organisms via probes or otherwise is known as
    (a) xenobiology (b) palaeobiology (c) femtobiology (d) exobiology

31. The branch of biology which deals with the effects of extraterrestrial space on living organisms such as the effects of outer space on man is known as
    (a) space biology (b) xenobiology (c) allobiology (d) femtobiology

32. The branch of biology that deals with the formation, structure and function of cells is known as
    (a) karyology (b) cytology (c) syncytiology (d) desmology

33. The study of cell nuclei, especially in reference to their chromosomes, is known as
    (a) pycnology (b) chromosomology (c) tychology (d) karyology

34. The branch of horticulture that deals with the study of fruits and their cultivation is known as
    (a) pomography (b) fruitology (c) pomology (d) economic botany

35. The study of tissues of the body is called
    (a) histology (b) cytography (c) histometry (d) dendrology

36. A special term has been coined to describe the science of the health of travellers. Name it.
    (a) Travellology (b) Sojournography (c) Emporiatrics (d) Heliology
37. There is a science which makes comparative study of the nervous system and systems of mechanical and electronic control of machinery in order to better understand communication and control in both types of systems. What is this science called?
    (a) Cybernetics (b) Cyclology (c) Neurocyclology (d) Biomechanical engineering
38. Phycology is the scientific study of
    (a) bacteria (b) algae (c) fungi (d) ferns
39. Ichthyology is the scientific study of
    (a) earthworms (b) frogs and other amphibians (c) hippopotamus and other mud-dwelling animals (d) fishes
40. What is the name given to the study of the behaviour of animals in their natural environment?
    (a) Psychoecology (b) Behaviourology (c) Ethology (d) Pocology

3
THE CELL

41. Which scientist was the first to discover the cell?
    (a) Regnier de Graaf (b) Anton Van Leeuwenhoek (c) Nehemiah Grew (d) Robert Hooke

42. In which structure did the abovenamed scientist see the cell?
    (a) Retina of a pig (b) Cork (c) Human skin (d) Sputum

43. What is the size of a typical plant or animal cell?
    (a) Less than five microns (b) Between five and forty microns (c) Between fifty and hundred microns (d) Between hundred and two hundred microns

44. A group of cells with similar structure and performing similar functions is called
    (a) tissue (b) cell group (c) syncytium (d) organism

45. In the centre of almost every cell, a big spherical body is seen. What is this known as?
    (a) Nucleolus (b) Endoplasmic reticulum (c) Nucleus (d) Ribosome

46. Who first propounded the cell theory stating that all living things are made up of cells or of material formed by cells?
    (a) Theodor Schwann (b) Matthias Jakob Schleiden (c) Asa Gray (d) Alfred Russel Wallace

47. Cell theory as it was first propounded was not able to explain how new cells are formed and its propounders were hard put to explain it. Who was the first biologist to explain that cells divide and that all new cells must come from pre-existing cells?

(a) Robert Brown (b) Thomas Young (c) Rudolf Virchow (d) Georges Cuvier

48. The cells in which the hereditary material or the DNA is not surrounded by a membrane are called
(a) procaryotic cells (b) membraneless cells (c) eucaryotic cells (d) cells without nuclei

49. The cells in which the hereditary material or the DNA is surrounded by a membrane to form a true nucleus are called
(a) nucleated cells (b) non-nucleated cells (c) eucaryotic cells (d) advanced cells

50. The average volume of a procaryotic cell is
(a) less than 0.1 μm^3 (b) between 0.2 μm^3 and 10 μm^3 (c) between 20 μm^3 and 50 μm^3 (d) between 50 μm^3 and 100 μm^3

51. The average volume of an eucaryotic cell is
(a) between 10 μm^3 and 100 μm^3 (b) between 100 μm^3 and 1,000 μm^3 (c) between 1,000 μm^3 and 10,000 μm^3 (d) generally greater than 10,000 μm^3

52. Powerhouses of the cells are
(a) ribosomes (b) endoplasmic reticulum (c) golgi body (d) mitochondria

53. An adhesive part of an epithelial cell by which it adheres to adjoining cells is known as
(a) cell membrane (b) cell wall (c) desmosome (d) plasma membrane

54. Mitosis is the name given to the process of cell division. What is the name given to the period between two mitotic cycles?
(a) Prophase (b) Interphase (c) Metaphase (d) Telophase

55. What are or were the Hela Cells?
(a) Cells taken from a rare animal called Hela
(b) Artificial cells produced by genetic engineering

(c) Cells from patients suffering from fatal disorders
(d) Cells from a black American woman who died in 1951

56. Which word related to the cell is derived from a Greek word meaning 'thread'?
    (a) Ribosomes (b) Golgi body (c) Mitosis (d) Pinocytosis

57. This is a complex, non-carbohydrate polymer found in cell walls. Its function is to provide stiffening to the cell as in xylem vessels and bark fibres. Name it.
    (a) Lignin (b) Saponin (c) Opsonin (d) Plant polypetide

58. Surprisingly, according to the current scientific thought, one of the cell organelles is thought to have descended from aerobic bacteria. Which one?
    (a) Ribosomes (b) Mitochondria (c) Lysosomes (d) Microsomes

59. Which of the following types of microscopy permits the direct observation of living cells?
    (a) Light microscopy (b) Transmission electron microscopy (c) Scanning electron microscopy (d) Phase-contrast microscopy

60. Pyknosis, karyorrhexis and karyolysis indicate that the cell is
    (a) resting (b) going to divide (c) dead (d) actively synthesizing the secretory granules

61. What does the mitotic index of a cell population or tissue tell us?
    (a) The proportion of cells which are dead (b) The proportion of cells which are undergoing mitosis (c) The proportion of cells which are at complete rest (d) The term does not exist

62. It is now widely accepted that the cell cycle is made up of four stages. What is the correct sequence of these stages?
    (a) M, G_1, S, G_2 (b) G_1, G_2, M, S (c) S, M, G_1, G_2 (d) G_1, M, G_2, S

63. The scientific term for the interval between consecutive phases of mitosis is
    (a) intermitosis (b) interphase (c) gap (d) intramitosis

64. Which of the following drugs can block mitosis?
    (a) Colchicine (b) Tetracycline (c) Chloromycetin (d) Paromomycin

65. The most remarkable feature of this particular cell organelle is that it contains surprisingly large quantities of an enzyme, acid hydrolase. Name it.
    (a) Golgi apparatus (b) Microsomes (c) Nucleosomes (d) Lysosomes

66- All body surfaces, whether internal or external, are
72. covered by a layer of cells known as epithelium whose main function is protection. Below are shown some major types of epithelia. Can you identify them?

66

67

68

69

70

71

72

73. Which of the following cells can be seen with a naked eye?

    (a) Cells of the elephant's brain (b) Neurons of a giant squid (c) Egg cells of a hen (d) No single cell can be seen with a naked eye.

74. Which of the following cell-organelles are also known as chondriosomes?

    (a) Ribosomes (b) Golgi body (c) Lysosomes (d) Mitochondria

75. This cell organelle is stained best with osmium tetroxide or silver nitrate. Name it.
    (a) Golgi body (b) Desmosomes (c) Nucleolus (d) Endoplasmic reticulum

76. Which cell organelle is absent from highly differentiated cells which have lost the power of division, such as nerve cells?
    (a) Nucleus (b) Centrosphere (c) Golgi body (d) Lysosomes

77. A white blood cell with a lobed nucleus and cytoplasmic granules that stain with acidic dyes is called
    (a) basophil (b) macrophage (c) eosinophil (d) thrombocyte

78. A white blood cell containing granules that stain with basic dyes is called
    (a) basophil (b) basic nucleocyte (c) basic W.B.C. (d) any of the above

79. A type of white blood cell with a very large nucleus, rich in DNA, and a small amount of clear cytoplasm, is found in the blood. It produces antibodies and is important in defence against disease. Name it.
    (a) Fibroblast (b) Myeloid cell (c) Lymphocyte (d) Neutrophil

80. A cell or organism containing only one representative from each of the pairs of homologous chromosomes found in the normal diploid cell is known as a
    (a) half cell (b) potential cell (c) haploid cell (d) tetraploid cell

81. During reproduction in certain fungi, when the two hyphae fuse together, there is fusion of cytoplasm, but the nuclei remain separate. What is this phenomenon known as?

(a) Plasmogamy (b) Karyogamy (c) Syncytum (d) Fibroblast

82. If a scientist wants to examine a cell under an electron microscope, which stain might be useful for him?
    (a) Saffranin (b) Eosin (c) Uranyl acetate (d) None

83. Which of the following cell organelles occurs in pairs at right angles to one another near one pole of the nucleus?
    (a) Golgi complex (b) Centriole (c) Mitochondria (d) Endoplasmic reticulum

84. In which of the following cells would you be able to find the centriole?
    (a) All animal cells (b) All plant cells (c) Motile algae (d) All bacteria

85. Which of the following cell organelles is sometimes referred to as dictyosome in plants?
    (a) Mitochondria (b) Golgi complex (c) Lysosomes (d) Ribosomes

86. Which cell organelles are known to cell biologists by the more colourful term 'suicide bags'?
    (a) Mitochondria (b) Desmosome (c) Smooth endoplasmic reticulum (d) Lysosomes

87. Which of these cell organelles are full of peroxide-destroying enzymes, catalases?
    (a) Rough endoplasmic reticulum (b) Desmosomes (c) Microbodies (d) Ribosomes

88. What is the name given to a large amoeboid cell that engulfs and destroys bacteria invading the human body?
    (a) Macrophage (b) Mast cell (c) Elastocyte (d) T-cell

89. Which tissue of the body forms a heat-insulating layer just beneath the skin?

(a) Muscle tissue (b) Adipose tissue (c) Areolar tissue (d) Tendons
90. Chondrocyte is another name for
    (a) bone cells (b) plasma cells (c) cartilage cells (d) muscle cells
91. Osteocyte is another name for
    (a) bone cell (b) muscle cell (c) retinal cell (d) intestinal cell
92. The diameter of human red blood cells is
    (a) 2 μm (b) 7 μm (c) 15 μm (d) 50 um

4
CHROMOSOMES AND GENES

93. The word chromosome is derived from Greek words meaning
    (a) thread-like body (b) coloured body (c) agent of inheritance (d) coiled snakes

94. A technique has been devised whereby treatment with a low-concentration salt solution swells the cells and disperses the chromosomes. They can then be photographed and the photograph can be cut into sections, each containing a separate chromosome. If these chromosomes are matched into pairs and then arranged in the order of decreasing length, the result is called
    (a) karyotype (b) karyosome (c) chromograph (d) genetic fingerprinting

95. Which of the following organisms is supposed to have the maximum number of chromosomes?
    (a) Whale (b) Mouse (c) Drosophilia (d) Adder's tongue fern

96. Certain living species such as wheat are referred to as polyploids. What are polyploid organisms?
    (a) Many organisms growing from a single cell (b) A plant or animal showing infinite variations (c) A plant or animal displaying more than two basic sets of chromosomes (d) An organism whose chromosome number changes from generation to generation

97. Which organism did the American zoologist Thomas Hunt Morgan choose for his genetic experiments?

(a) Garden peas (b) Drosophilia (c) Cockroaches (d) Frogs

98. This famous scientist worked with a mold called *Neurospora crassa* and came to the conclusion that the characteristic function of the gene was to supervise the formation of a particular enzyme. Who was he?

    (a) George Wells Beadle (b) William Clouser Boyd (c) John Joseph Bittner (d) Wendell Meredith Stanley

99. What does the term gene pool mean?

    (a) Total number of micro-organisms found in a swimming pool (b) Harmful micro-organisms in a pool (c) Total number of different kinds of genes pooled by all the members of a population (d) Total number of genes in all cells of a single individual

100. How many plant species are polyploids?

    (a) About hundred (b) One-tenth of all plant species (c) One-seventh of all plant species (d) One-third of all plant species

101. Active DNA is called

    (a) euchromatin (b) heterochromatin (c) metachromatin (d) telochromatin

102. When stained with a certain dye, the chromosomes show alternate light and dark bands known as Q-bands under the microscope. Name the dye.

    (a) Giemsa stain (b) Haematoxylin (c) Eosin (d) Quinacrine mustard

103. The position of a gene at a particular point on a chromosome is called

    (a) position (b) posting (c) locus (d) allele

104. What are alleles?

    (a) Defective genes (b) Two genes occupying the same position on a pair of chromosomes (c) Genes which are unable to express themselves (d) Genes which are present in the mitochondria

105. When both genes of an allelic pair are partially and almost equally expressed, inheritance is said to be
(a) heterozygous (b) semi-inheritance (c) intermediate (d) partial

106. What is pleiotropy?
(a) The appearance of two or more characters which are controlled by a single gene (b) Multiple genes producing one character only (c) One gene expressing in different ways in different individuals (d) Mutation of a gene which passes on to offsprings

107. The expression of one gene sometimes prevents the expression of another gene at a different chromosome. What is this phenomenon called?
(a) Telosuppression (b) Genetic inhibition (c) Polyploidism (d) Epistasis

108. Sex chromatin was first noticed in
(a) nerve cells of female cats (b) ovaries of human female (c) buccal mucosa of female athletes (d) hair roots of male horses

109. In some chromosomes, there are small terminal regions which are attached to the rest of the chromosome by a narrow, non-staining region. What are these terminal regions known as?
(a) Peripheral bodies (b) Chromatids (c) Satellites (d) Chromatophores

110. In how many groups are the 23 human chromosome pairs divided?
(a) Five (b) Seven (c) Nine (d) Eleven

111. What is the Philadelphia chromosome?
(a) An extra chromosome found in the residents of Philadelphia (b) An abnormal chromosome found in the patients of certain blood cancers (c) A chro-

mosome common between a gorilla and a man
(d) A hypothetical chromosome thought to have been present in the Neanderthal man

112. Nature has an inherent mechanism whereby harmful genes are eliminated from the population by causing the death of the bearer of such harmful genes. Modern medicine is able to save these individuals, thereby causing the accumulation of harmful genes. What is this phenomenon known as?
(a) Genetic drift (b) Genetic involution (c) Genetic devolution (d) Genetic pollution

113. A particular branch of genetics has been called bean-bag genetics. Name it.
   (a) Population genetics (b) Animal genetics
   (c) Genetics related to a family of bean plants
   (d) Genetics pursued by amateurs

114. It is now well known that genes contain coding sequences for proteins. However, in most genes coding sequences are interrupted by non-coding sequences. What are these non-coding sequences called?
   (a) Exons (b) Introns (c) Variable regions
   (d) Microsatellites

115. Any DNA or RNA fragment which has the ability to become reversibly incorporated into the chromosome of a cell in order to replicate is known as
   (a) mutational gene (b) foreign gene (c) jumping gene (d) codon

116. A collection of closely linked genes that tend to behave as a single unit is known as
   (a) supergene (b) genome (c) polygene (d) giant gene

117. Which of the following reagents stains DNA and is particularly used to show chromosomes during cell division?
    (a) Methylene blue (b) Iodine (c) Borax Carmine (d) Feulgen's stain

118. A group of three nucleotide bases that codes for a specific amino acid is known as
    (a) an anticodon (b) a codon (c) mRNA (d) tRNA

119. What is the width of the DNA helix?
    (a) 2 nm (b) 10 nm (c) 12 nm (d) 25 nm

120. The DNA helix makes one full turn every
    (a) 3.4 nm (b) 7.8 nm (c) 20 nm (d) 30 nm

121. DNA found in the nucleus is wrapped around a simple protein which is basic in nature. This protein is also thought to regulate DNA functioning in some way. Name this protein.
    (a) Papain (b) Renin (c) Collagen (d) Histones

122. What is a nucleosome?
    (a) An extra nucleus found in a cell (b) A small nucleus (c) A structural unit composed of four pairs of histone proteins around which DNA is wrapped (d) A structural unit composed of 6 RNA molecules

5
THE PROTEINS

123. The word protein comes from
 (a) a Latin word meaning 'healthy' (b) a Greek word meaning 'of first importance' (c) a Latin word meaning 'chemical of life' (d) a Greek word meaning 'costly chemical'

124. A complex protein substance produced in living cells that accelerates chemical reactions in an organism without being permanently changed itself is known as
 (a) enzyme (b) co-factor (c) haemoglobin (d) diastase

125. Proteins on heating change from
 (a) liquid to gas (b) solid to liquid (c) solid to gas directly without changing first into liquid (d) liquid to solid

126. Proteins are made up of several
 (a) fatty acids (b) lactose molecules (c) amino-acids (d) different molecules all containing phosphorus

127. Which was the first enzyme to be prepared from animal tissue?
 (a) Pancreozymin (b) Pepsin (c) Amylase (d) Enterogastrin

128. What does the following chemical formula stand for?

(a) Amino acid (b) Fatty acid (c) Peptide (d) Simple protein

129. How many amino acids are known to occur in nature?
(a) Only 20 (b) About 50 (c) Over 200 (d) Infinite

130. What is common among the following amino acids: phenylalanine, leucine, methionine and valine?
(a) They are not required by humans (b) They are essential amio acids (c) They are the only amino acids not found in milk (d) Females do not require these animo acids

131. What is a complete protein?
(a) A protein which contains all twenty amino acids (b) A protein which contains all essential amino acids (c) A protein which contains fatty acid and carbohydrate residues (d) A protein which does not solidify on heating

132. Where would you normally find the protein *fibroin*?
(a) Milk (b) Wool (c) Silk (d) Cartilage

133. Approximately how many different types of proteins are there in the human body?
(a) 5,000 (b) 25,000 (c) 100,000 (d) 1,000,000

134. There is a region on every antigen molecule that is unique to it and is therefore responsible for the antigen's specificity in the antigen-antibody reaction. What is this region known as?
(a) Gamma region (b) Lyophilic region (c) Complement (d) Epitope

135. Biologists frequently have to test material for the presence of proteins. For this, they use a mixture of mercury, nitric acid and water. What is this reagent known as?
(a) Millon's reagent (b) Frederick's reagent (c) Proprotein regeant (d) Protein reagent

136. Which of the following substances are aptly known as biological catalysts?
    (a) Hormones (b) Antibiotics (c) Enzymes (d) DNA
137. Which of the following enzymes is used for coagulating milk protein to obtain casein?
    (a) Trypsin (b) Ptyalin (c) Rennin (d) Pancreatin
138. What is an apoenzyme?
    (a) An appropriate enzyme (b) An enzyme found only in plants (c) An enzyme having a metallic atom (d) The protein part of an enzyme
139. Which is the most abundant of all the proteins in the higher vertebrates?
    (a) Protamine (b) Collagen (c) Elastin (d) Reticulin
140. Who was the first scientist to work out the exact sequence of amino acids in various protein molecules?
    (a) Frederick Sanger (b) Jean Dausset (c) Hugo de Vries (d) Arthur Kornberg
141. Which was the first protein hormone ever to be synthesized?
    (a) Insulin (b) Glucagon (c) Oxytocin (d) Testosterone

6
VITAMINS

142. Which of the following scientists coined the word 'vitamin'?
 (a) Casimir Funk (b) Linus Pauling (c) Carl F. Cori (d) Paul Ehrlich

143. Which of the following animals don't need vitamin C?
 (a) Monkeys (b) Guinea pigs (c) Bulbuls (d) Cats

144-153. Vitamin B is a complex group of several vitamins. That is why this group of vitamins is often known as Vitamin B Complex. The vitamins of this group are often known by numbers. Can you match the numbers with the correct vitamin?

144. B_1 (a) Niacin
145. B_2 (b) Pyridoxine
146. B_3 (c) Cyanocobalamin
147. B_5 (d) Thiamine
148. B_6 (e) Adenylic acid
149. B_8 (f) Folic acid
150. B_{12} (g) Riboflavin
151. B_c (h) Paraaminobenzoic acid
152. B_t (i) Pantothenic acid
153. B_x (j) Carnitine

154. What is the other name for biotin?
 (a) Vitamin H (b) Vitamin P (c) Vitamin K (d) Vitamin D

155-159. Deficiencies of many vitamins of the B-complex group give rise to specific diseases. Can you match deficiencies of vitamins with the disease they produce?

155. B_1 (a) Angular Stomatitis

156. B_2 (b) Megaloblastic anaemia
157. B_3 (c) Peripheral neuritis
158. B_6 (d) Pellagra
159. B_{12} (e) Beri Beri

160- Many vitamins are known not only by the alphabet
164. of the English language but by other names as well, depending upon the disease they prevent. Can you match the vitamins here?

160. Vitamin K   (a) Antiinfection vitamin
161. Vitamin A   (b) Antirachitic vitamin
162. Vitamin D   (c) Antihaemorrhagic vitamin
163. Vitamin C   (d) Antisterility vitamin
164. Vitamin E   (e) Antiscorbutic vitamin

165- Can you match the following vitamins with their dis-
169. coverers?

165. Vitamin A   (a) Edward Mellanby
166. Vitamin C   (b) Herbert McLean Evans
167. Vitamin D   (c) Stefan Ansbacher
168. Vitamin E   (d) Szent-Gyorgyi
169. Vitamin K   (e) Elmer Verner McCollum
                            Marguerite Davis

170. Which vitamin is also known as PP vitamin?
    (a) Vitamin A (b) Niacin (c) Vitamin D (d) Pantothenic acid

171. Which one of the following vitamins was once known as Vitamin G?
    (a) Riboflavin (b) Biotin (c) Paraaminobenzoic acid (d) Coagulation factor

172. Which vitamin is also known as Tocoferol?
    (a) Vitamin A (b) Vitamin B (c) Vitamin D (d) Vitamin E

173. The deficiency of which of the following vitamins can cause night-blindness?
    (a) A (b) B (c) C (d) D

7
THE ORIGIN OF LIFE

174. The theory that living things can be produced only by other living things, and cannot develop spontaneously from non-living materials is known as
    (a) abiogenesis (b) non-spontaneous creation (c) biogenesis (d) biochemical evolution

175. He believed in the theory of spontaneous generation and even divulged, partly, a method for the production of human beings from non-living materials! Who was he?
    (a) Paracelsus (b) Wilhelm Von Waldeyer (c) Galen (d) Avicenna

176. An interesting theory regarding the origin of life on earth suggests that primitive life forms could have reached the earth from elsewhere in the universe — either 'planted' deliberately by other intelligent beings or else brought accidentally by meteorites. What is this theory popularly known as?
    (a) Berlitz's theory (b) Panspermia hypothesis (c) Theory of exotic creation (d) Ex-earth biogenesis

177. For a proper appreciation of origin of life on earth, a few facts about earth itself are necessary. How old is our earth?
    (a) 4.6 trillion years (b) 4.6 billion years (c) 4.6 million years (d) 4.6 thousand years

178. For most of the time during the history of earth all the continents remained fused together as Pangaea. When did the continents start drifting away?
    (a) 2 billion years ago (b) 1 billion years ago (c) 500 million years ago (d) 200 million years ago.

8
MICRO-ORGANISMS

Bacteria

179. Bacteria occur in various shapes and sizes. The ones shaped like a rod are called
 (a) bacilli (b) cocci (c) chlamydomonas (d) spirilli

180. Bacteria which are spherical are called
 (a) neisseria (b) balleria (c) cocci (d) spirilli

181. Bacteria which are spiral in shape are called
 (a) springeria (b) spirilli (c) balleria (d) cork-screw bacteria

182. Certain bacteria are able to thrive under extreme environmental conditions such as absence of oxygen, high salt concentration, high temperature or highly acidic environment. What are these bacteria called?
 (a) Hardy bacteria (b) Neobacteria (c) Probacteria (d) Archaebacteria

183. Halophiles are the bacteria which live in
 (a) extremely hot springs (b) extremely strong brine (c) arctic areas (d) marshy places

184. Bacteria which are found in marshy areas and among the flora of cattle rumen are called
 (a) methanogens (b) thermophiles (c) acidophiles (d) thermoacidophiles

185. Many bacteria possess fine, whip-like parts which they use for swimming. What are these parts called?
 (a) Spirilla (b) Flagella (c) Coculla (d) Whipulla

186. Which scientist in 1884 developed a valuable method of staining the bacteria which rendered them visible under the microscope?

(a) Sir Wilfred Thomason Grenfell (b) Major Greenwood (c) Hans Christian Joachim Gram (d) James Graham

187. Which disease is caused by the bacterium *Treponema pallidum?*
(a) Gonorrhoea (b) Syphilis (c) Hepatitis (d) Conjunctivitis.

188. Which family of bacteria can decompose a wide variety of organic compounds, natural or man-made, including inert ones like hydrocarbons? This property has actually been used by scientists to reduce pollution such as that from petroleum spillage.
(a) Enterobacteria (b) Spirochaetes (c) Archaebacteria (d) Pseudomonads

189. In the beginning there was no oxygen in earth's atmosphere. The activity of which organism is generally supposed to have made the earth's atmosphere aerobic or oxygen-rich?
(a) Cyanobacteria (b) Filamentous algae (c) Monocotyledons (d) Mycoplasma

190. Which non-toxic, fast growing cyanobacterium is cultivated in tanks as a protein-rich animal feed?
(a) *Nostoc* (b) *Spirulina* (c) *Stigonema* (d) *Oscillatoria*

191. *Leptothrix* is an example of iron bacteria. Which of the following statements best fit the iron bacteria?
(a) They cause rusting of iron (b) They are responsible for preservation of iron-rich artefacts in archaeological material (c) They oxidise ferrous compounds to ferric oxide (d) They cause iron to be deposited in the livers of animals.

192. Many bacterial cells contain an extra-chromosomal genetic element which replicates independently of

the chromosomal DNA. What is this genetic element known as?

(a) Plerome (b) Plasmin (c) Plectostele (d) Plasmid

193. A plasmid found in some bacteria can integrate reversibly with the chromosome of its host and then it behaves as part of the host chromosome, multiplying with it. What is this plasmid known as?

(a) Episome (b) Exosome (c) Extrosome (d) Introsome

194-
196. Below you find three types of micro-organisms and the optimum temperatures they need for growth. Can you make a proper match?

Organism	Optimum temperature
194. Psychrophilic	(a) 60°C
195. Mesophilic	(b) 120°C
196. Thermophilic	(c) 25°C-45°C

Viruses

197. Who first discovered viruses?
(a) George Beadle (b) Martinus Willem Beijerinck (c) Lorenz Oken (d) Anton Van Leeuwenhoek

198. The term 'virus' means
(a) an infinitesimally small particle (b) violent (c) killer (d) poison

199. Capsid is the
(a) RNA strand of the virus (b) rudimentary cytoplasm within the virus (c) protein coat of the virus (d) hair-like process seen on the surface of viruses

200. What is the name given to the process when the viral nucleic acid is incorporated into a bacterial chromosome to become an integral part of it?
(a) Lysogeny (b) Reverse transcription (c) Transformation (d) Transduction

201. Bacteriophages are
    (a) bacteria which kill viruses (b) a colony of bacteria and viruses living together (c) Viruses which infect bacteria (d) bacteria which can change into viruses and vice-versa

202. The transfer of a gene or genetic material from one bacterial cell to another by a virus is known as
    (a) transposition (b) transduction (c) translocation (d) genetic drift

203. Which scientist was the first to discover the existence of viruses that can infect bacteria?
    (a) John Kendrew (b) Dimitri Ivanovski (c) Alan Hodgkin (d) Felix d'Herelle

204- Here you can find pictures of some common viruses.
211. How many of them can you identify? They are drawn on the same scale, so size should also give a clue. Identify also the diseases they cause.

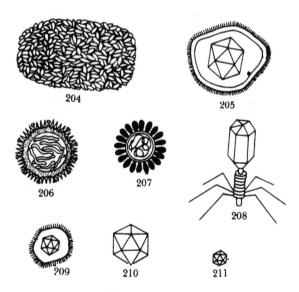

212. An RNA-containing virus whose genome becomes integrated into the host DNA by means of the enzyme reverse transcriptase and then replicates with it is known as
(a) retrovirus (b) bacteriophage (c) virion (d) thermovirus

213. When was the first AIDS case discovered in India?
(a) 1982 (b) 1985 (c) 1986 (d) 1987

Other microorganisms

214. Like viruses, these organisms live inside the cells, but like bacteria they multiply by binary fission. One of the diseases caused by them is trachoma. Name these organisms.
(a) Virions (b) Vireids (c) Chlamydiae (d) Mycoplasma

215. These are a group of very small intra-cellular bacteria. They are transmitted by arthropod vectors. Some of the diseases caused by them are typhus, trench fever and Q fever. Name these organisms.
(a) Rickettsia (b) Bacteroids (c) Viroids (d) Mycoplasma

9
HEREDITY AND EVOLUTION

General

216. Reappearance of an inherited trait after an apparent gap of several generations is called
    (a) jumping traits (b) atavism (c) autotrophic traits (d) hidden traits

217. Which English writer bitterly attacking Darwin's theories remarked, 'A hen is only an egg's way of producing another egg'?
    (a) Samuel Butler (b) Herbert George Wells (c) Thomas Hardy (d) Thomas Henry Huxley

218. Who was the first top-ranking biologist to devise a scheme rationalizing the evolutionary development of life, and maintaining that the species were not fixed but that they changed and developed?
    (a) Erasmus Darwin (b) Carolus Linnaeus (c) Georges Cuvier (d) Jean Baptiste Lamarck

219. Who wrote the book *Zoological Philosophy* in 1809?
    (a) Charles Darwin (b) Jean Baptiste Lamarck (c) Carolus Linnaeus (d) Anton Van Leeuwenhoek

220. The occurrence of two forms among the organisms of the same kind is known as
    (a) diastrophism (b) diatropism (c) dimorphism (d) dichroism

221. A group of closely related, structurally and functionally similar organisms which interbreed with one another in nature, but not with organisms of other groups is known as
    (a) species (b) genus (c) order (d) family

222. What is common among the following pairs of animals: mallard and the pintail duck; tiger and the lion; the platy and swordtail fish; polar bear and the Alaskan brown bear?

(a) They are members of same species who have acquired different characters (b) They are members of different species which can interbreed under captivity (c) They are species which are ecologically dependent upon each other (d) They are animals which compete with each other for food

223. Which of the following is the best evidence that gymnosperms have descended from pteridophytes?

(a) They are evergreen (b) They have cones (c) Some gymnosperms have ciliated sperms (d) Some gymnosperms cannot bear fruit

224. Development of particular characteristics in a small group of creatures through chance rather than by natural selection is known as

(a) indeterminism (b) genetic isolation (c) chance evolution (d) genetic drift

225. The independent evolution of geographically separate but related species in a way that produces similar forms is known as

(a) convergent evolution (b) co-evolution (c) parallel evolution (d) synergistic evolution

226. Inheritance in which maternal characters are shown by male offspring and paternal by female is known as

(a) Alternative inheritance (b) criss-cross inheritance (c) Blending inheritance (d) amphi-gonous inheritance

227. The type of inheritance in which a character appears in all the female offspring but in none of the males is known as

(a) hologynic inheritance (b) maternal inheritance (c) holandric inheritance (d) alternative inheritance

228. In 1831, Charles Darwin began his historic five-year cruise round the world, during which he collected data for his theory of evolution. What was the name of the ship?
(a) HMS Discovery (b) HMS Nature (c) HMS Explorer (d) HMS Beagle

229. Identify this skull

230. The development of similar structures in unrelated organisms as a result of living in similar ecological conditions is called
(a) convergent evolution (b) parallel evolution (c) devolution (d) retrograde evolution

231. When man purposely selects desirable breeds of plants and animals for further propagation, the process is known as
(a) natural selection (b) unnatural selection (c) artificial selection (d) superselection

Human evolution

232. Discussing the search for human origins, who made this famous remark, 'Scientists have already cast

much darkness on this subject, and if their investigations continue, we shall soon know nothing at all'?

(a) Sir Alexander Fleming (b) Mark Twain (c) George Bernard Shaw (d) Oscar Wilde

233. Which of the following hominids were the first to use fire?

(a) *Australopithecus boisei* (b) *Homo habilis* (c) *Homo erectus* (d) *Proconsul*

234. This famous scientist was a believer in Darwinism. When the Bishop of Oxford, Samuel Wilberforce, asked him sarcastically if he traced his own descent from the apes through his father or mother, he answered disdainfully, 'If I had to choose as an ancestor either a miserable ape or an educated man who could introduce such a remark into a serious scientific discussion, I would choose the ape.' Name this witty scientist.

(a) Thomas Henry Huxley (b) Theodor Schwann (c) Alfred Russel Wallace (d) Claude Bernard

235. Which famous person, when asked to choose between apes and angels as the forebears of man, replied, 'I am on the side of the angels'?

(a) Oscar Wilde (b) Sir Arthur Conan Doyle (c) Ernest Hemingway (d) Benjamin Disraeli

236. Which scientist independently arrived at the same conclusions regarding evolution as did Darwin?

(a) Ernst Heinrich Haeckel (b) Alfred Russel Wallace (c) Justus Von Liebig (d) Gregor J. Mendel

237. Which great discovery regarding human evolution was made in 1857?

(a) Radiocarbon dating was perfected (b) Geological periods were demarcated for the first time (c) Bones of Neanderthal man were discovered

(d) Importance of vestigial structures was recognised for the first time

238. He was a great zoologist, yet so opposed was he to the notion of evolution through natural selection that he wrote anonymous articles against it and even fed opponents with anti-evolutionary facts so that they could create trouble in the scientific world. Identify this scientist.
(a) Sir Richard Owen (b) Matthias Jakob Schleiden (c) Theodor Schwann (d) Asa Gray

239. Which is the oldest hominid species yet found?
(a) *Australopithecus boisei* (b) *Australopithecus robustus* (c) *Australopithecus africanus* (d) *Australopithecus afarensis*

240. Which important discovery did Mary Leakey make in 1948?
(a) She exposed the Piltdown hoax (b) She discovered the skull of *Proconsul* (c) She discovered that radioactivity could be used to age fossils (d) She discovered a new method for rapid discovery of fossils

241. Who discovered the remains of 'Lucy', the most famous australopithecine found in Ethiopia in 1974?
(a) Richard Leakey (b) Louis Leakey (c) Donald Johanson (d) F. Clark Howell

242. The idea that man evolved from ape-like ancestors was not generously received. One of the most famous remarks made was, "My dear, descended from apes! Let us hope it is not true, but if it is, let us pray that it will not become generally known." Who made this interesting remark?
(a) Wife of the Bishop of Worcester (b) Conservative British Prime Minister Benjamin Disraeli (c) William Ewart Gladstone (d) American President Andrew Johnson

10
DIVERSITY OF ANIMAL LIFE

General

243. How many species of living animals have been catalogued by biologists so far?
 (a) 5,000 (b) 50,000 (c) 120,000 (d) 1,200,000

244. How many species of animals did Carolus Linnaeus list in his book *Systema Naturae* (1758)?
 (a) 4,200 (b) 10,500 (c) 15,700 (d) Over 50,000

245. Which of the following phyla contains the largest number of animal species?
 (a) Chordata (b) Arthropoda (c) Mollusca (d) Echinodermata

246. In which of the following respects do mammals differ from reptiles?
 (a) Ability to breed on land (b) Arrangement of the legs (c) Structure of the teeth (d) Brighter colours

247. Which of the following animals belong to the same phylum as man?
 (a) Sea squirts (b) Grasshoppers (c) Fish (d) Starfish

248. Which of the following early animal characteristics led to the development of the body cavity?
 (a) Bilateral symmetry (b) Development of the mesoderm (c) Segmentation (d) Development of the liver within the abdomen

249. What is a Mexican hairless?
 (a) The name of a hairless bird which lives in Mexico (b) Another name for Gila Monster (c) Mexican kangaroo, a very rare species (d) A hairless breed of dog

250. What is the name given to organisms that are able to tolerate wide variations of salt concentrations?
    (a) Euryhaline (b) Stenohaline (c) Eutheria (d) Eusporangiate

251. There are certain special nerves which are insulated by myelin sheaths around them. The nerve impulse travels very fast along these nerves because it "leaps" along certain special points on these nerves. What is this special mode of conduction known as?
    (a) Ultrasonic conduction (b) Saltatory conduction (c) Superconduction (d) Hyperconduction

252. The outermost layer of the epidermis of vertebrates is known as
    (a) stratum lucidum (b) stratum germinativum (c) exodermis (d) stratum corneum

253. After passage of an electrical impulse along a nerve, there is a short period when no stimulus, however large, will evoke a further impulse. What is this period known as?
    (a) Resting period (b) Recovery period (c) Refractory period (d) Terminatory period

254. The arrangement of parts in an organism in such a way that the structure can only be divided into similar halves along one plane is known as
    (a) bilateral symmetry (b) radial symmetry (c) Uniplanar symmetry (d) unique symmetry

255. Which is the smallest warm-blooded animal in the world?
    (a) Pygmy shrew (b) Koala (c) Bee humming bird (d) Red ants

256. Which are the earliest animals ever to possess a true voice?
    (a) Fishes (b) Reptiles (c) Birds (d) Frogs

257. The simplest creatures that possess specialized nerve cells are
(a) annelids (b) coelenterates (c) pisces (d) porifera

258. Quadriped and biped are familiar terms but what is a palmiped?
(a) Animal which walks on its palms (b) Animal having five fingers in the shape of a palm leaf (c) Any bird with webbed feet (d) Animal such as kangaroo, who uses its tail in locomotion

259. If an animal is described as feral, what does this tell about it?
(a) A furred animal (b) A nocturnal animal (c) A domestic animal (d) A domestic animal gone wild

260. Large White, Wessex Saddleback and Landrace are three breeds of an animal native to Britain. Which one?
(a) Pig (b) Horse (c) Rabbit (d) Dog

261. Can you give another name for the beast of prey sometimes called a glutton?
(a) Puma (b) Polar bear (c) Wolverine (d) Leopard

262. The arrangement of parts in an organism in such a way that cutting in any plane across the diametre splits it into similar halves, is an example of
(a) radial symmetry (b) bilateral symmetry (c) natural symmetry (d) perpendicular symmetry

Prehistoric animals

263. Which is the smallest dinosaur yet found?
(a) Brachiosaurus (b) Tyrannosaurus (c) Compsognathus (d) Brachiosaur

264. How many species of extinct animals have been identified by fossil study?
(a) 10,000 (b) 50,000 (c) 100,000 (d) 500,000

265. Which fossil organism is usually regarded as the connecting link between reptiles and birds?
    (a) *Archaeopteryx* (b) *Tyrannosaurus rex* (c) Brontosaurus (d) Brachiosaurus

266. Which was the first dinosaur to be scientifically described?
    (a) Brachiosaurus (b) Diplodocus (c) Stegosaurus (d) Megalosaurus

267. The largest living terrestrial mammal is the African elephant. Which is the largest terrestrial mammal ever to roam the earth?
    (a) *Gigantophis garstini* (b) *Osteodontornis orri* (c) *Baluchitherium grangeri* (d) *Megaloceros giganteus*

268. Which was the heaviest of all prehistoric animals?
    (a) Iguanodon (b) Brachiosaurus (c) Stegosaurus (d) Tyrannosaurus

269. Identify this strange-looking bird?

    (a) Dodo (b) Sand Grouse (c) Skua (d) Sheathbill

270. Who was the first scientist to classify fossils?
    (a) William Smith (b) Georges Cuvier (c) Friedrich Wilhelm Humboldt (d) Antoina Laurent de Jussieu

271. The figure below shows the reconstructed skeleton of a prehistoric animal. Identify it.

Animals without backbones

272. Who is usually considered to be the founder of modern entomology?
    (a) Jan Swammerdam (b) Richard Lower (c) Olof Rudbeck (d) Marcello Malpighi

273. Which animal is generally considered as a connecting link between the annelid worms and the arthropods?
    (a) Barnacle (b) Fairy shrimp (c) Peripatus (d) Mantis shrimp

274. These animals depict a radial symmetry, i.e., their body is organized in a circle around a central axis. They have two layers of cells — ectoderm and endoderm — with a jelly-like substance between them. In which phylum are these animals grouped?
    (a) Platyhelminthes (b) Cnidaria (c) Mollusca (d) Annelida

275. What is Aristotle's lantern?
    (a) A community of starfishes living together (b) A group of fireflies living together in a tree and

glowing (c) Special equipment designed by Aristotle with which sea creatures could be found in ocean deep darkness (d) A special arrangement of teeth in sea urchins

276-280. Mouth parts of insects are modified according to their feeding habits. How many insects can you identify here from their mouth parts?

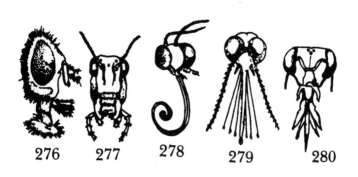

276   277   278   279   280

281. All the following cnidarians live in the sea except one. Name the exception.
(a) Corals (b) Sea anemones (c) Jellyfish (d) Hydra

282. Which are the most primitive multi-cellular animals?
(a) Soft corals (b) Sea anemones (c) Sponges (d) Starfishes

283. Which are the simplest animals to have a nervous system and a gut?
(a) Cnidarians (b) Molluscs (c) Echinodermates (d) Annelids

284. Which are the simplest three-layered animals alive today?
(a) Lampshells (b) Thorny-headed worms (c) Flatworms (d) Earthworms

285. Which of the following worms are also known as the proboscis worms?
    (a) Flatworms (b) Ribbon worms (c) Round worms (d) Horseshoe worms

286. The name of which phylum is derived from a Greek word meaning sea nymph?
    (a) Platyhelminthes (b) Nemertinea (c) Entoprocta (d) Mollusca

287. The gut of cnidarians has only one opening which serves both as mouth and anus. Which are the simplest animals to have a gut with two separate openings?
    (a) Rotifers (b) Hairworms (c) Roundworms (d) Horseshoe worms

288. Which of the following are also known as Gordian worms?
    (a) Hairworms (b) Tapeworms (c) Round worms (d) Pinworms

289. Which simple animal has helped the scientists to deduce that there were 400 days in a year, 300 million years ago?
    (a) Hydra (b) Paramecium (c) Coral (d) Sea anemone

290. What are the soft internal partitions of a coral polyp or sea anemone called?
    (a) Septa (b) Mesenteries (c) Theca (d) Coelome

291. What are the stinging cells on the tentacles of corals, sea anemones, jellyfish and hydra called?
    (a) Barbs (b) Statocysts (c) Nematocysts (d) Rays

292. Corals are best known for the massive 'rocky' reefs they build in the tropics. Which is the largest such reef made by them?
    (a) South Atlantic reef (b) Polynesian reef (c) Great Indian Ocean reef (d) Great Barrier Reef

293. Why are 'soft' corals so colourful?

(a) Because minute plants live in their tissues (b) Because of coloured rods or spines in their tissues (c) Because of pigment cells in their body walls (d) Because they manufacture many coloured chemicals

294. Which insects, common in Britain, are red with black spots, black with red spots or yellow with black spots?

(a) Stag beetles (b) Scarabs (c) Caddis flies (d) Ladybirds

295. What is an imago?

(a) The final adult stage in the life-cycle of an insect (b) Another name for a butterfly caterpillar (c) An insect produced by mating between two unrelated species (d) A sterile insect

296. The production of sounds by some insects by rubbing together parts of the body, usually to attract a female, is known as

(a) Cacophony (b) Stridulation (c) Undulation (d) Swan Song

297. There is a layer of jelly-like material that separates the ectoderm and endoderm in coelenterates. What is this layer called?

(a) Gelplasm (b) Mesoderm (c) Mesoglea (d) Mesocotyl

298. Which is the longest insect in the world?

(a) Tropical stick insect (b) Cicada (c) Tarantula (d) Spring-tails

299. Which are generally considered to be the longest-lived insects?

(a) Common housefly (b) Queen termites (c) Cicadas (d) Spring-tails

300. Which are generally considered to be the loudest of all insects?
(a) Goliath beetle (b) Hercules beetle (c) Male cicadas (d) Desert locusts

301-322. Below you will find some major phyla of invertebrates (animals without backbones). How many animals can you match with their phyla?

Phylum		Animals
301. Porifera	(a)	Roundworms
302. Ctenophora	(b)	Rotifers
303. Platyhelminthes	(c)	Lamp shells
304. Nematoda	(d)	Echiurid worms
305. Nematomorpha	(e)	Water bears
306. Priapulida	(f)	Starfishes
307. Rotifera	(g)	Arrow worms
308. Gastrotricha	(h)	Priapulid worms
309. Acanthocephala	(i)	Phoronid worms
310. Ectoprocta	(j)	Soft-bodied molluscs
311. Phoronida	(k)	Peanut worms
312. Brachiopoda	(l)	Beard worms
313. Annelida	(m)	Tongue worms
314. Sipunculoidea	(n)	Acorn worms
315. Echiuroidea	(o)	Flatworms
316. Pentastomida	(p)	Moss animals
317. Tardigrada	(q)	Earthworms
318. Mollusca	(r)	Comb jellies
319. Chaetognatha	(s)	Horsehair worms
320. Pogonophora	(t)	Hairy backs
321. Echinodermata	(u)	Sponges
322. Hemichordata	(v)	Thorny-headed worms

323- Below are three types of insects with their description. Make a proper match.
325.

Insect type	Features
323. Apterygota	(a) Pass through 3 stages of development
324. Exopterygote	(b) Pass through 4 stages of development
325. Endopterygote	(c) Having no wings

326- Of all forms of life, the most varied are the insects
343. (*class insecta*). More than 700,000 species are known. Class *insecta* is divided into several orders. Below are several orders and common names of insects. Make a proper match.

Orders	Common names of insects
326. Diplura	(a) Mayflies
327. Thysanura	(b) Stick insects
328. Collembola	(c) Termites
329. Odonata	(d) Bugs
330. Ephemeroptera	(e) Book lice
331. Plecoptera	(f) Cockroaches
332. Mantodea	(g) Earwigs
333. Phasmida	(h) Bird lice
334. Orthoptera	(i) Web-spinners
335. Blattaria	(j) Sucking lice
336. Isoptera	(k) Thrips
337. Dermaptera	(l) Silverfish and relatives
338. Embioptera	(m) Dragonflies
339. Psocoptera	(n) Two-pronged bristletails
340. Mallophaga	(o) Praying mantises
341. Anoplura	(p) Springtails
342. Hemiptera	(q) Grasshoppers and relatives
343. Thysanoptera	(r) Stoneflies

344. Animals such as barnacles, crabs, lobsters, shrimps and woodlice are grouped in the class

(a) Arachnida (b) Crustacea (c) Onychophora (d) Chilopoda

345-354. Below you will find some more orders of the class *insecta* along with their common names. Match them.

Orders	Common names of insects
345. Neuroptera	(a) Scorpionflies
346. Megaloptera	(b) True flies
347. Raphidiodea	(c) Butterflies
348. Mecoptera	(d) Fleas
349. Trichoptera	(e) Snakeflies
350. Coleoptera	(f) Bees, wasps and ants
351. Lepidoptera	(g) Alderflies
352. Diptera	(h) Beetles and weevils
353. Siphonaptera	(i) Caddis flies
354. Hymenoptera	(j) Lacewings and ant lions

Fishes

355. Which of the following periods is usually known as the age of fishes?
    (a) Ordovician period (b) Silurian period (c) Devonian period (d) Permian period

356. Approximately how many species of fishes are known to biologists?
    (a) 100,000 (b) 75,000 (c) 30,000 (d) 5,000

357. Which group of fish is the most abundant, diverse and complex?
    (a) Bony fishes (b) Cartilaginous fishes (c) Jawless fishes (d) Fishes with bony plates

358. What is the staple diet of a basking shark?
    (a) Plankton (b) Invertebrates living on the sea bed (c) Other fish (d) Leftovers thrown by man

359. Most sharks are heavier than water. How do they prevent themselves from sinking?

(a) By means of a swim bladder (b) By movement through the water (c) By expelling water through the gills (d) By extracting hydrogen from water and using it for buoyancy

360. What does this figure show?

361. The spiracle, an opening behind the eye leading to the gill system, is the characteristic of
    (a) lampreys (b) hagfishes (c) cartilaginous fishes (d) bony fishes

362. To which of the following classes do the rays belong?
    (a) Cyclostomes (b) Elasmobranchs (c) Bradyodonts (d) Osteichthyes

363. Which of the following senses is most important to a shark when hunting its prey?
    (a) Hearing (b) Sight (c) Smell (d) Touch

364. Which of the following fishes was thought to have been extinct until its recent discovery?

(a) Garpike (b) Bowfin (c) Coelacanth (d) Hagfish

365. What is the swim bladder in a bony fish used for?
(a) Respiration (b) Fast swimming (c) Buoyancy (d) Collection of urine during swimming

366. Pelagic fish live
(a) near the top of the sea (b) near the bottom of a lake (c) in the middle of a pond (d) near the bottom of a river

367. The order of bony fish that contains the lungfish is
(a) elopiformes (b) dipnoi (c) clupeiformes (d) myctophiformes

368- Below you will find some common bony fishes and
378. their orders. Can you make a proper match?

Orders	Common names of fishes
368. Notacanthiformes	(a) Toadfishes
369. Anguilliformes	(b) Cods and relatives
370. Osteoglossiformes	(c) Milk fish
371. Myctophiformes	(d) Angler fishes
372. Gonorhynchiformes	(e) Whalefishes and squirrel fishes
373. Batrachoidiformes	(f) Spiny eels
374. Lophiiformes	(g) Snakeheads
375. Gadiformes	(h) Bony tongues
376. Beryciformes	(i) Flying fishes
377. Atheriniformes	(j) Lantern fish
378. Channiformes	(k) Eels

Ambhibians

379. How many million years ago did the first amphibian appear?
(a) 1000 (b) 780 (c) 350 (d) 120

380. How many living species of amphibians are known to biologists?
(a) Over 50,000 (b) 12,780 (c) 10,000 (d) 2,300

381. Which is the largest amphibian in the world?
    (a) Chinese giant salamander (b) Axolotl (c) Great siren (d) Great crested newt

382. To which of the following groups do the newts and salamanders belong?
    (a) Anura (b) Apoda (c) Urodela (d) Gastrotheca

383. What is the main food of young tadpoles?
    (a) Small aquatic invertebrates (b) Algae and other plant material (c) Other tadpoles (d) Leftovers thrown by man

384. Which of the following amphibians is not normally seen in its adult form?
    (a) Hairy frog (b) Ceylonese caecilian (c) Axolotl (d) Spadefoot

385. In which of the following group of amphibians do the young larvae feed in the oviduct on 'uterine milk' consisting of oil droplets and sloughed-off cells from the oviduct?
    (a) Newts (b) Salamanders (c) Toads (d) Caecilians

386. With which animal are the caecilians often confused?
    (a) Earthworms (b) Starfishes (c) Salamanders (d) Centipedes

387. How do caecilians find their way through the earth and locate their prey?
    (a) By sight (b) By sensory tentacles (c) By smell (d) By chemical means

388. Which of the following genuses of frogs secrete toxic skin secretions which local Indians use as arrow poison?
    (a) *Pelobates* (b) *Xenopus* (c) *Dendrobates* (d) *Gastrotheca*

389. The young of which of the following anurans develop in honeycomb depressions on the mother's back?

(a) *Rana temporaria* (b) *Bufo bufo* (c) *Pipa pipa* (d) *Gigantorana goliath*

390. Which of the following groups of Anurans have a digging tubercle on the undersurface of the hindfoot which is used to excavate a burrow?

(a) Spadefoots (b) Clawed toads (c) Tree frogs (d) Bullfrogs

391. There are about 2,900 species of frogs in the world. How many of them are sea-dwellers?

(a) About half of them (b) About one-fifth of them (c) Only about 100 species (d) None

Reptiles

392. How many million years ago did the first reptiles appear on earth?

(a) 2,000 (b) 1,000 (c) 300 (d) 125

393. How many living species of reptiles are known to biologists?

(a) 50,000 (b) 20,000 (c) 11,300 (d) 6,000

394. Class *Reptilia* is divided in four major orders. To which order do tortoises and turtles belong?

(a) *Chelonia* (b) *Squamata* (c) *Rhynchocephalia* (d) *Crocodilia*

395. To which order do snakes and lizards belong?

(a) *Chelonia* (b) *Squamata* (c) *Rhynchocephalia* (d) *Subreptilia*

396. Which of the following snakes kills its prey by constriction?

(a) Viper (b) Python (c) Banded krait (d) Rat snake

397. Which of the following snakes has a fang which can be swivelled to lie flat in the mouth when not in use?
    (a) Cobra (b) Boa (c) Coral Snake (d) Pit Viper
398. Which of the following reptiles give birth to live young?
    (a) Sea snakes (b) Pythons (c) Slowworms (d) All of them
399. The shell of a turtle or tortoise is divided into an upper and a lower section. What is the lower section called?
    (a) Carapace (b) Scute (c) Plastron (d) Subshell
400. What is the largest living member of the crocodile family?
    (a) Estuarine crocodile (b) American alligator (c) Nile crocodile (d) Gharial
401. Which is the only snake to build a nest?
    (a) Tree snake (b) Rat snake (c) Python (d) King Cobra
402. Which of the following reptiles can fly?
    (a) Draco (b) Fringed gecko (c) Golden tree snake (d) Moloch
403. This lizard living in deserts of Australia has a curious ability to use rare rain water. The rain water does not pass through the skin but flows in minute channels along it until it reaches the mouth where it is swallowed! Name this lizard.
    (a) Rhinoceros iguana (b) Moloch (c) Bearded lizard (d) Teiid lizard
404. Which is the longest venomous snake in the world?
    (a) Anaconda (b) Boa constrictor (c) King Cobra (d) Reticulated python
405. How does the male snake locate the female for breeding?
    (a) By smell (b) By sight (c) By looking for her trail marks (d) By hearing her movements

Birds

406. How many million years ago did the birds evolve from their reptilian ancestors?
 (a) 570 (b) 430 (c) 320 (d) 150

407. How many species of birds have been catalogued so far?
 (a) 100,000 (b) 72,000 (c) 46,800 (d) 8,650

408. Which is the largest living bird?
 (a) Brown Kiwi (b) Ostrich (c) Emperor Penguin (d) Greater Rhea

409. Which is the heaviest flying bird?
 (a) Kori bustard (b) Sunbittern (c) Bensch's Rail (d) Kagu

410-419. The feet of various birds show adaptation to their habits. How many birds can you identify from their feet?

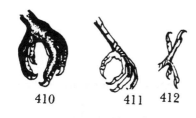

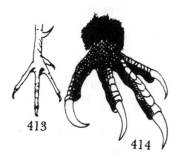

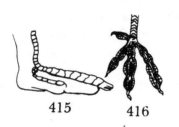

415   416

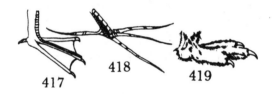

417   418   419

420. Which is generally reputed to be the smallest bird in the world?

    (a) Waxwing (b) Sickle-billed vanga (c) Helena's hummingbird (d) Red-vented bulbul

421. Which bird has the highest energy consumption per unit of weight?

    (a) Kagu (b) Hummingbird (c) Bustard (d) Ostrich

422. Generally flightless birds are very heavy, which is one reason they can't fly. Which is the smallest flightless bird?

    (a) Flightless rail (b) Great-crested Grebe (c) Black-throated Diver (d) Fulmar

423. Which is the fastest-flying bird?
    (a) Arctic tern (b) Yellow warbler (c) Spine-tailed swift (d) Apapane
424. The red colour of birds like flamingos is produced by
    (a) pigment zooerythrin (b) pigment melanin (c) simple reflection (d) small crystals of mercuric oxide in its plumage.
425- The bills of different types of birds have become
434. adapted to their feeding habits. Identify the following birds from their bills?

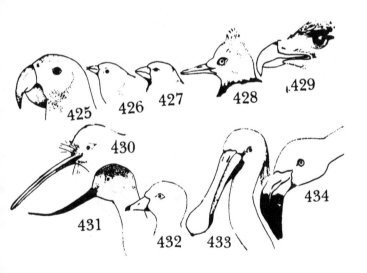

435. Which of the following birds has colonies in Alaska and Greenland, both groups migrating each year to winter in Africa but by different routes?
(a) Wheatear (b) Swallow (c) Wandering albatross (d) Orange-bellied flowerpecker

436. Which of the following features is characteristic of birds but not reptiles?
(a) Presence of a pygostyle (b) Separate ilium, ischium and pubis bones (c) Hollow bones (d) Presence of ribs

437. Both reptiles and birds share a number of anatomical features which led biologists to believe that birds descended from reptiles. What are those features?
(a) Presence of only one occipital condyle (b) Lower jaw made up of five parts (c) Presence of an egg tooth in the young at the time of hatching (d) Presence of uncinate process on the ribs.

438- Many birds have characteristic nests and can indeed
446. be identified by looking at their nests. How many birds can you identify here?

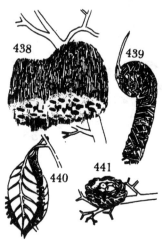

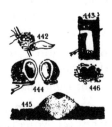

447. Birds evolved from reptiles and hence earliest fossils of birds reveal that they had teeth. Which is the oldest known specimen of a toothless bird?
(a) Archaeopteryx (b) Archaeornis (c) Ichthyornis (d) Hesperornis

448. The tongue of this bird is sticky and has a hooked tip as shown, which helps it to retrieve insects from depths. Identify this bird.

(a) Barbet (b) Honey guide (c) Toucan (d) Woodpecker

449. Penguins are found in
    (a) Antarctic ice fields only (b) Arctic region only
    (c) Both Arctic and Antarctic regions (d) Both Polar regions as well as in many forests

450. If a bird is said to be 'raptorial', what kind is it?
    (a) Land bird (b) Water bird (c) Bird of prey
    (d) Bird which lives in symbiosis with a mammal

451. What is the name given to the song box of a bird?
    (a) Nares (b) Trachea (c) Syrinx (d) Lungs

452. What are a bird's wing flight feathers called?
    (a) Remiges (b) Rectrices (c) Plumulae (d) Down

453. What is the main function of the bastard wing or alula?
    (a) To control turbulence and prevent stalling
    (b) To provide lift (c) To act as a brake when landing (d) To provide acceleration

454. The main flight muscle of the birds, the *major pectoral*, is anchored at one end to the breast bone. To which wing bone is it attached on the other side?
    (a) Ulna (b) Radius (c) Humerus (d) Femur

455. Which of the following birds is unusual in moulting twice a year?
    (a) Cuckoo (b) Lap/wing (c) Willow warbler
    (d) Woodpecker

456. In birds, what is the uncinate process?
    (a) A freely moveable bone on which the lower jaw is hinged (b) An outgrowth of the rib which strengthens the rib cage (c) A bony knob at the back of the skull (d) The remnant of the egg tooth which is used by the young one to break the egg

457. Which is the heaviest bird ever to have existed?
    (a) Moa (b) Ostrich (c) Elephant bird (d) Kiwi

458- Below are some common birds and their orders.
481. Make the proper match?

Orders	Common names of birds
458. Passeriformes	(a) Owls
459. Piciformes	(b) Cuckoos
460. Coraciformes	(c) Game birds
461. Coliformes	(d) Nightjars and relatives
462. Apodiformes	(e) Birds of prey
463. Caprimulgiformes	(f) Albatrosses and relatives
464. Strigiformes	(g) Perching birds
465. Cuculiformes	(h) Storks
466. Psittaciformes	(i) Pelicans
467. Columbiformes	(j) Swifts and relatives
468. Charadriiformes	(k) Penguins
469. Gruiformes	(l) Woodpeckers and relatives
470. Galli/formes	(m) The Ostrich
471. Falconiformes	(n) Cranes
472. Anseriformes	(o) Grebes
473. Ciconiiformes	(p) Kingfishers and relatives
474. Pelecaniformes	(q) Emu and Cassowaries
475. Procellariiformes	(r) Ducks
476. Sphenisciformes	(s) Gulls
477. Gaviformes	(t) Kiwis
478. Podicipediformes	(u) Mousebirds
479. Casuariiformes	(v) Pigeons
480. Struthioniformes	(w) Divers
481. Apterygiformes	(x) Parrots

Mammals

482. The highest sound frequency which a human being can detect is 20,000 vibrations/second. What is the highest frequency which a bat can detect?
    (a) 10,000 (b) 50,000 (c) 75,000 (d) 150,000

483. The word mammal comes from a

(a) Chinese word meaning hairy (b) Latin word meaning breast (c) Greek word meaning warm-blooded animal (d) German word meaning beautiful

484. Which mammal has a natural sun-burn lotion?
    (a) Reindeer (b) Elephant (c) Giraffe (d) Hippopotamus

485- Many mammals can be identified from their toes.
489. Identify the mammals here.

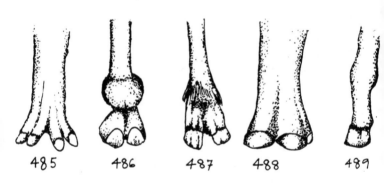

490. Mammals are divided into several orders. Which order includes the most advanced mammals?
    (a) Pholidota (b) Sirenia (c) Primates (d) Cetacea

491. In which order are the whales, dolphins and porpoises included?

(a) Cetacea (b) Tupaioidea (c) Dermoptera (d) Pinnipedia

492- This diagram depicts the shapes of feet and hands
497. of six primates. Identify them.

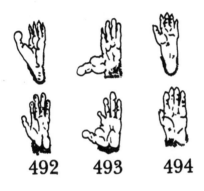

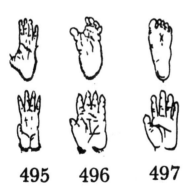

498- By studying the teeth of mammals, a good idea can
500. be formed about their eating habits. What are the eating habits of these three mammals?

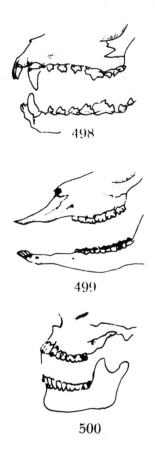

501. Polar bears live at
(a) North Pole only (b) South Pole only (c) both North and South Poles (d) both Polar regions as well as in several forests

502. How many cavities does the stomach of a cow have?
    (a) Two (b) Four (c) Eight (d) Nine

503. By and large animals belonging to different species don't mate together, but occasionally animals of different species can mate in captivity. Which two animals are known to mate together in captivity?
    (a) Cats and dogs (b) Camels and giraffes (c) Chimpanzee and gorillas (d) Lions and tigers

504. The order of mammals that contains the even-toed ungulates, in which the weight of the body is supported on the third and fourth digits only, is
    (a) artiodactyla (b) perissodactyla (c) cetacea (d) mursupialia

505. Which order of mammals contains the bats?
    (a) Chelonia (b) Insectivora (c) Chiroptera (d) Edentata

506. Which biologist was the first to show that almost all the Australian mammals were marsupials?
    (a) Charles Darwin (b) Sir Joseph Banks (c) Friedrich Wilhelm Humboldt (d) Franz Joseph Gall

507. When explorers brought the first skins of this mammal from Australia, zoologists thought it was a hoax. It is also known as water mole or duck mole. Name this mammal.
    (a) Spiny anteater (b) Sloth (c) Wombat (d) Duckbill Platypus

508. The pouch in which a Kangaroo mother keeps her young baby is called
    (a) Marsupium (b) Uterus (c) External placenta (d) Cloaca

509. Which order of mammals is the most primitive of the placental mammals?
    (a) Chiroptera (b) Edentata (c) Insectivora (d) Pholidota

510. Which rare mammal has been slaughtered by villagers because it is feared as a harbinger of death?
(a) Gibbon (b) Lesser Mouse Lemur (c) Macaque (d) Aye Aye

511. There are about 900 species of bats in the world. Of these, which well-known species feeds on the fresh blood of warm-blooded vertebrates such as man?
(a) Horseshoe bats (b) Flying foxes (c) Vampires (d) Mouse-tailed bats

512-527. Below you will find some major orders of class *mammalia* (mammals). Match the animal with its proper order.

Order		Animals
512. Monotremata	(a)	Gnawing mammals
513. Marsupialia	(b)	Hoofed mammals with odd number of toes
514. Insectivora	(c)	Elephants
515. Primates	(d)	Egg Laying mammals
516. Edentata	(e)	Hyraxes
517. Pholidota	(f)	Sea cows
518. Dermoptera	(g)	Pangolins
519. Rodentia	(h)	Aardvark
520. Lagomorpha	(i)	Pouched mammals
521. Carnivora	(j)	Flying Lemurs
522. Pinnipedia	(k)	Anteaters, armadillos and sloths
523. Perissodactyla	(l)	Flesh-eating mammals
524. Sirenia	(m)	Lemurs and tarsiers
525. Tubulidentata	(n)	Seals and Walrus
526. Hyracoidae	(o)	Rabbits, hares and pikas
527. Proboscidae	(p)	Insect and worm-eating mammals

11
DIVERSITY OF PLANT LIFE

General

528. How many species of plants have been catalogued by biologists so far?
 (a) 5,000 (b) 50,000 (c) 500,000 (d) 50,000,000

529. How many species of plants did Carolus Linnaeus list in his book *Species Plantarum* (1753)?
 (a) 5,900 (b) 7,900 (c) 10,900 (d) 15,700

530. Kelps are a kind of
 (a) fungi (b) red algae (c) brown algae (d) green algae

531. Which thickening agent widely used in food, textile and pharmaceutical industry is extracted from kelps?
 (a) Fucic acid (b) Mucilage (c) Cellulose (d) Alginic acid

532. Several algae are held to the bottom of a pond or sea by a structure known as
 (a) stipe (b) holdfast (c) lamina (d) root

533. Which of the following land plants has descended from red algae?
 (a) Tomato (b) Rose (c) All bryophytes (d) None

534. *Sargassum* is a kind of
 (a) brown algae (b) blue green algae (c) green algae (d) red algae

535. Which of the following pigments are found in the red algae?
 (a) Phycoerythrin (b) Phycocyanin (c) Carotene (d) Fucoxanthin

536. Which of the following algae is the deepest growing algae in the seas?

(a) Red algae (b) Blue algae (c) Blue green algae (d) Green algae

537. Agar agar is a gelatinous substance widely used by microbiologists to prepare their culture media. From which of the following is this substance extracted?

(a) Fungi (b) Green algae (c) Red algae (d) Brown algae

538. Which of the following pigments is present in brown algae?

(a) Phycobilins (b) Fucoxanthin (c) Anthocyanin (d) biliverdin

539. What is the name given to plants which can not grow on soil rich in time or calcium carbonate?

(a) Cacti (b) Dye Plants (c) Calcifuge (d) Gooseberries

540. On the fur of which animal does green algae usually grow, tinting its fur green?

(a) Armadillo (b) Rabbits (c) Bats (d) Sloths

541. For which of the following reasons are the green algae believed by biologists to be ancestors of land plants?

(a) They contain chlorophyl (b) They have a cellulosic cell wall (c) They store starch (d) All of the above

542. Except tree ferns, most ferns have rhizomes. What are rhizomes?

(a) Special roots (b) Underground stems (c) Bipinnate leaves (d) Primitive flowers

543. After a forest fire when above-ground plants are killed, which plants are the first to reappear?

(a) Ferns (b) Mosses (c) Liverwosts (d) Angiosperms

544. Certain special cells having highly thickened walls, known as stone cells, are found in
    (a) stones (b) shells of nuts (c) pulp of guava (d) pulp of pear

545. Which plant tissue is mainly concerned with the transport of water and minerals?
    (a) Phloem (b) Cambium (c) Xylem (d) Sclereids

546. Through which tissue of the plant is food and other organic substances translocated?
    (a) Phloem (b) Metaxylem (c) Cambium (d) Sclereids

547. The way in which leaves are arranged on a stem is known as
    (a) apostasis (b) phyllotaxy (c) epitopy (d) heterophylly

548. The wood produced by gymnosperms is commercially known as
    (a) logwood (b) sapwood (c) hardwood (d) softwood

549. Botanically speaking, which of the following is a false fruit?
    (a) Tomato (b) Pea (c) Coconut (d) Apple

550. The wall of a true fruit is known as
    (a) Exocarp (b) Epicarp (c) Pericarp (d) Mesocarp

551. What are parthenocarpic fruits?
    (a) Very hard fruits (b) Fruits formed without fertilization (c) Fruits enclosed in pods such as pea (d) Fruits with more than one seed at their centre

552. Pineapple is
    (a) a simple fruit (b) a composite fruit (c) an aggregate fruit (d) a parthenocarpic fruit

553. Strawberry is

(a) an aggregate fruit (b) a false fruit (c) a parthenocarpic fruit (d) a simple fruit

554. Which plant has the distinction of bearing the biggest seed?
(a) Tambalacoque tree (b) African baobab (c) Mangroves (d) Coco de mer

555. With a growth rate of about 5 mm a minute, it is one of the fastest growing organisms in the world. It is also known as 'the lady of the white veil'. Identify it.
(a) Bamboo (b) Stinkhorn fungus (c) Sphagnum moss (d) Clubmoss

556- Pollen grains of many plants and trees have quite
565. characteristic shapes when viewed under a microscope. Identify the trees from their pollens.

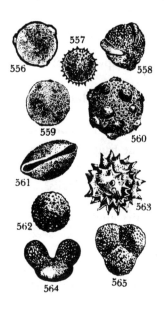

566- Leaves come in many different shapes. How many
596. shapes can you identify here?

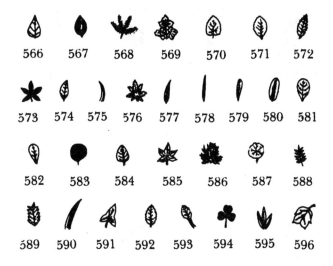

597- Arrangement of leaves on the stem varies greatly.
600. How many arrangements are here?

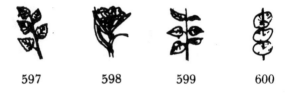

| 597 | 598 | 599 | 600 |

601- Some flowers such as roses, daffodils and tulips grow
611. singly from the tip of a flower stalk, but the majority of blossoms grow in clusters known as inflorescences. Identify the types of inflorescences here.

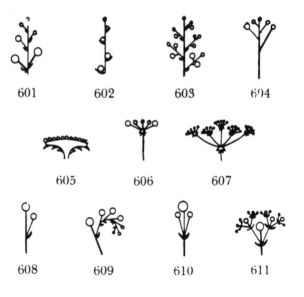

612. The embryo enclosed within the seed consists of two parts — the radicle and the plumule. What is the significance of plumule?
(a) It gives rise to root system (b) It gives rise to the shoot (c) It stores glycogen (d) It is mainly a protective covering of the embryo.

613. A root system composed mainly of branches rather than of one principal root is known as
(a) fibrous root system (b) tap root system (c) adventitious roots (d) branched root system

614. What are the taproots?
(a) Roots which emerge around water taps (b) Roots of weeds growing around accumulated water (c) A root system in which there is one major root with other roots branching off it (d) A root system in which roots arise from stems

615-619. In some plants tap roots are modified for storing reserve materials. Match the shape of the root with the plant.

615.	Fusiform	(a)	Turnip
616.	Napiform	(b)	Carrot
617.	Conical	(c)	Dahlia
618.	Tuberous	(d)	Radish
619.	Fasciculated	(e)	Sweet potato

620. Adventitious roots are
(a) week roots (b) roots arising in places other than the root system (c) roots which store foods (d) roots which are used as a fuel

621. Higher plants such as trees reproduce by seeds while lower plants such as mosses and ferns reproduce by spores. What is the basic difference between a spore and a seed?
(a) Spores take a much longer time to germinate (b) Spores lack a food reserve (c) Spores are easily

destroyed by animals (d) Spores need great nutrition from the parent plant to develop

622. Which are the earliest plants to have a conduction system for conduction of water and food?
(a) Mosses (b) Liverworts (c) Ferns (d) Gymnosperms

623. Which are the earliest plants to bear seeds?
(a) Mosses (b) Ferns (c) Gymnosperms (d) Angiosperms

624. Which of the following characterstics is distinctive of gymnosperms?
(a) They are evergreen (b) They have cones (c) They grow on hilly areas only (d) Their eggs are in direct contact with the air

625. What is the name of the plant whose berries are used to make gin?
(a) *Juniperus* (b) *Cupressus* (c) *Chamaecyparis* (d) *Cedrus*

626. Most Bryophytes such as mosses and liverworts thrive only in moist places. Why?
(a) They derive energy by splitting water (b) They have no conduction system to transport water (c) Water is needed to cool down their cells (d) Bryophytes have some special cells called hydrocytes which need water

627. The region of a plant in which active cell division (mitosis) occurs is known as
(a) mericarp (b) meristele (c) meristem (d) sclereid cells

628. The cylinder of tissues lying inside the endodermis of plant stems and roots, containing the vascular tissues, is known as
(a) stele (b) cambium (c) pith (d) pericycle

629. The layer of plant cells between the endodermis and the phloem, consisting mainly of parenchyma, from which lateral roots originate, is known as
(a) cambium (b) pith (c) pericycle (d) vascular bundles

630. If the parenchyma cells contain chloroplasts then the tissue comes to be called as
(a) chlorenchyma (b) aerenchyma (c) collenchyma (d) sclerenchyma

631. Cork-like tissue with large air-filled cavities between cells, present in the stems and roots of certain water plants, is known as
(a) collenchyma (b) tracheids (c) aerenchyma (d) casparian bands

632. In this type of plant tissue, the cells are similar to parenchyma but have cellulose wall thickenings, particularly at the angles. What is this tissue known as?
(a) Procambium (b) Lateral meristem (c) Intercalary meristem (d) Collenchyma

633. In pteridophytes and gymnosperms the xylem tissue consists mainly of
(a) Vessel members (b) Tracheids (c) Tracheae (d) Sclerenchyma

634. Which of the following statements about monocotyledons are true?
(a) Cambium is absent (b) Vascular bundles are scattered randomly (c) Vascular bundles are arranged in a single ring (d) Cambium is present in the centre of the stem

635. The centre of the plant stem consists of parenchyma cells and is called the
(a) endarch (b) pith (c) protoxylem (d) meta-xylem

636- Below are shown some major types of vascular bun-
641. dles, encountered in various plants. Can you give
the name of each arrangement?

Key:

Phloem   Xylem   Cambium

634   635   636

637   638   639

642- One finds several kinds of adventitious roots in na-
646. ture. Can you match the kind of adventitious root
with the plant?

642. Beaded roots   (a) Banyan tree
643. Prop roots   (b) Tinospora
644. Stilt roots   (c) Cuscuta
645. Assimilatory roots   (d) Sugarcane
646. Haustoria   (e) Bitter gourd

647. Small plants with a soft stem are called
(a) shrubs (b) herbs (c) vires (d) grasses

648. Medium-sized plants with woody stems that branch
profusely from the base and attain a bushy appearance are called
(a) bushes (b) vires (c) weeds (d) shrubs

649. The concept totipotency means that
(a) every plant feels pain (b) plants can communicate among one another (c) every living plant cell should be able to regenerate a whole plant (d) any plant is potent enough to grow in the vicinity of other plants

650. This large brown seaweed was once an important source of potash, iodine and soda. Name it.
(a) sargassum (b) kelp (c) leek (d) lichens

651. Plants living perched on other plants but not obtaining water or food from them are called
(a) epiphytes (b) periphytes (c) hydrophytes (d) epinastic plants

652. *Chlamydomonas* reproduces asexually by forming
(a) gametophores (b) zoospores (c) gametophytes (d) ciliated cells

653. A plant that grows in soils with a high concentration of salt, as found in salt marshes, is known as a
(a) marinophyte (b) halophyte (c) aquaphyte (d) periphyte

654. Any specialized plant cell that is dispersed among cells of a different kind is called a
(a) normoblast (b) xenophyte (c) macrophyll (d) idioblast

655. In certain flowers, the anthers show dehiscence lines towards the centre of the flower, thus favouring self-pollination. What are these anthers called?
(a) Introrse (b) Extrorse (c) Inverted anthers (d) Centranthers

656. The inner layer of the cell wall surrounding the pollen grains of angiosperms and gymnosperms is known as
(a) exine (b) intine (c) integument (d) involucre

657. In certain plants, the alternating generations of their life-cycle are morphologically identical. What is this condition known as?
(a) Isomorphism (b) Heteromosphism (c) Isograft (d) Twinning of generations

658. Plants grown in total darkness lack chlorophyll and thus appear white or yellow. What is this type of growth known as?
(a) Stuntism (b) Heterotrophic growth (c) Etiolation (d) Euploid growth

659. An ovary or fruit having separate or distinct carpels is known as
(a) syncarp (b) apocarp (c) heterocarp (d) homocarp

660. Plant cells thought to be gravity sensitive are known as
(a) gravitons (b) gravitocytes (c) statoliths (d) cytomeres

661. The stalk of the flower is called
(a) receptacle (b) pedicel (c) thalamus (d) stigma

662. Each stamen of a flower may be regarded as a highly modified
(a) stem (b) root (c) flower (d) leaf

663. The young bryophyte gametophyte that develops following spore germination is known as
(a) protonema (b) protostele (c) protoderm (d) sporangium

664. Which of the following statements is true about bryophytes?
(a) The gametophyte is diploid (b) The sporophyte is haploid (c) The sporophyte is dependent on gametophyte for support, shelter and nutrition (d) The gametophyte is dependent on sporophyte for support, shelter and nutrition

665. The female sex organ of the bryophytes, pteridophytes and most gymnosperms is known as
(a) archegonium (b) antheridium (c) protovule (d) ovule

666. The fern plant is
(a) haploid (b) diploid (c) tetraploid (d) euploid

667. During certain months of the year, small brown or yellow spots can be seen on the underside or margin of fern leaves. These spots are known as
(a) gametes (b) placenta (c) somites (d) sori

668. The covering that encloses the developing sporangia in the sorus of a fern is known as
(a) infundibulum (b) osculum (c) indusium (d) placenta

669. When the spore of a fern plant falls on moist ground, it germinates to form a green filament. It soon grows into a flat, green, heart-shaped structure known as
(a) sporophyte (b) gametophyte (c) adventitious thallus (d) prothallus

670. *Chilgozas*, the common dry fruit eaten in many Indian families, are the seeds of a
(a) pine (b) angiosperm (c) tree fern (d) daisy plant

671. The production of two different sizes of spores in some ferns is known as
(a) homospory (b) heterospory (c) dispory (d) polyspory

672. Which stain is used for demonstrating lignin?
(a) Methylene blue (b) Leishman's stain (c) Aniline blue (d) Phloroglucinol

673. Which stain is used for demonstrating fungal hyphae and spores?

(a) Aniline blue (b) Aniline sulphate (c) Aniline hydrochloride (d) Iodine

674. The fruits of which tree contain a sweet, edible dark brown pulp, sometimes called St. John's bread?
(a) Century plant (b) Carob (c) Maidenhair tree (d) Loquat

675. Colourless plastids are known as
(a) chloroplasts (b) chromoplasts (c) leucoplasts (d) coloplasts

676- Tropism is a directional growth movement of part
680. of a plant in response to an external stimulus. Below are given several types of stimuli and the type of tropism involved. Make a proper match.

Stimulus	*Type of tropism*
676. Light	(a) Thigmotropism
677. Gravity	(b) Hydrotropism
678. Chemical	(c) Geotropism
679. Water	(d) Chemotropism
680. Solid surface (touch)	(e) Phototropism

Fungi

681. Fungi are termed parasites when they assimilate tissues of living plants or animals. What are they termed when they live on decaying plant or animal matter?
(a) Saprobes (b) Holorobes (c) Phytogens (d) Telophytes

682. On many occasions, fungi have even altered the course of history. Which fungus reduced the population of Ireland from eight million in 1845 to six million a decade later?
(a) *Pencillium glaucum* (b) *Phytophthora* (c) *Synchytrium* (d) *Olpidium*

683. In which major group of fungi is the common black bread mould placed?

(a) Basidiomycetes (b) Ascomycetes (c) Zygomycetes (d) Sporomycetes

684. In which group are yeasts, cup fungi and edible morels placed?
    (a) Ascomycetes (b) Zygomycetes (c) Sporomycetes (d) Basidiomycetes

685. In which group are mushrooms, toadstools, puff balls and bracket fungi placed?
    (a) Ascomycetes (b) Zygomycetes (c) Sporo-mycetes (d) Basidiomycetes

686. Which of the following fungus causes the well-known leaf-curl disease of peaches, apricots and almonds?
    (a) *Taphrina deformans* (b) *Fusarium graminiarum* (c) *Saccharomyces cerevisiae* (d) *Trichoderma viride*

687. Many fungi display the phenomenon of heterothallism. What exactly is heterothallism?
    (a) Presence of a net-like mycelium (b) Production of different kinds of spores (c) Presence of different types of hyphae acting as male or female (d) Ability to produce both sexually as well as asexually

688. Fungi in which only the asexual method of reproduction has been observed are called
    (a) perfect fungi (b) fungi imperfecti (c) vegetative fungi (d) true fungi

689. Approximately how many species of fungi are known to biologists?
    (a) Over 200,000 (b) About 100,000 (c) 50,000 (d) 20,000

690. Many fungi contain in their cell walls a substance which is not present in higher plants. Name it.
    (a) Latex (b) Glycogen (c) Chitin (d) Ribose sugars

691. Mitosis in many fungi differs in a remarkable way from that observed in plant or animal cells. What is that?

(a) Nuclear envelope may not disappear (b) Cell swells to four times just before division (c) Cell develops affinity for a yellow dye, picric acid (d) Cell becomes hydrophobic

692. Which fungus is economically important in the brewing and baking industries?

(a) *Erysiphe graminicola* (b) *Sclerospora graminicola* (c) *Gibberella fujikuroi* (d) *Saccharomyces cerevisiae*

693. Which fungus is grown for its high protein content and is used as an additive to human and animal foods?

(a) *Fusarium Oxysporum* (b) *Fusarium graminiarum* (c) *Sclerotinia fruticola* (d) *Ustilago tritici*

694. Which fungus produces an enzyme which breaks cellulose into its constituent glucose molecules and thus can literally be used to convert old newspapers into a rich sugar syrup?

(a) *Trichoderma viride* (b) *Ustilago tritici* (c) *Gibberella fujikuroi* (d) *Sclerospora graminicola*

695. It was a fungus which gave one of the greatest drugs to the world of medicine. Identify it.

(a) *Penicillium notatum* (b) *Ustilago maydis* (c) *Colletorichum falcatum* (d) *Alternaria solani*

696. The symbiotic association of a fungus with the root of a higher plant is called

(a) mycosis (b) mycophagy (c) mycorrhiza (d) mycophilic association

697. In what way does a fungus growing around the roots of higher plants benefit them?

(a) It helps in absorbing water, nitrogen and other minerals (b) It produces growth-promoting substances (c) It produces antimicrobial substances (d) All of them

698. Spore-bearing organs of mushrooms hang down under the cap. They are usually referred to as

    (a) *fungus virilus* (b) plates (c) sieves (d) gills

699. The largest part of a fungus is the network of narrow, tubular branches which remains unseen beneath the surface of the soil. What is it called?

    (a) Mycelium (b) Ascus (c) Sporangium (d) Stalk

700. Which of the following fungi can reproduce both asexually and sexually?

    (a) Mucor (b) *Penicillium glaucum* (c) *Penicillium notatum* (d) Both (a) and (b)

701. Which famous person was so fond of mushrooms that he made it illegal for anyone else to eat those growing in the papal states, lest there should be a shortage for his own table?

    (a) Gregory XI (b) Louis XIV (c) Charles the Bold (d) Clement VII

702. Which of the following fungi is also known as the kerosene fungus?

    (a) Birch bracket (b) *Amorphotheca resinae* (c) *Myriostoma Coliformis* (d) Common morel

703. Truffles are rare delicacies much sought after by chefs all over the world. To which genus do they belong?

    (a) *Tuber* (b) *Pythium* (c) *Mutinus* (d) *Anthurus*

704. Truffles grow underground and have to be smelt out to locate them. Which animal is used most to smell out truffles?

    (a) Dog (b) Pig (c) Reindeer (d) Goat

705. Which insects cultivate fungus in their colonies and allow only the queen, king, and young ones to eat it?

    (a) Bees (b) Termites (c) Butterflies (d) Beetles

12
GREAT EXPERIMENTS

706. In 1668, he prepared eight flasks with meat inside. He sealed four while leaving other four open. Maggots developed only in open jars, disproving the theory of spontaneous generation. Who was this brilliant biologist?

    (a) Andreas Vesalius (b) Francesco Redi (c) Georg Ernst Stahl (d) Anton Van Leeuwenhoek

707. It is known only by its laboratory number 178, and is produced by U.S. scientists at Ohio University. What is it?

    (a) An underwater breathing apparatus which can be taken to twice the depth achieveable previously (b) A special germ-free chamber where organ transplant victims are kept (c) A small elephant which is as big as dog and thus could prove to be an interesting pet animal (d) A mouse two and half times larger than a normal mouse

708. He tied off an artery or a vein in a living animal to see on which side of this blockage the pressure within the blood vessel would build up. Thus he discovered circulation. Who was he?

    (a) William Harvey (b) Praxagoras of Cos (c) Andreas Vesalius (d) Thomas Hunt Morgan

709. In 1836, he prepared extracts of the glandular lining of the stomach and treated it with mercuric chloride, causing the enzyme pepsin to precipitate. This was a major breakthrough in the understanding of the process of digestion. Who was the biologist involved?

(a) Matthias Jakob Schleiden (b) Karl Nageli (c) Theodor Schwann (d) Karl Theodor Ernst Siebold

710. In 1752, he placed meat in small metal cylinders open at both ends — the ends being covered by wire gauze — and persuaded a hawk to swallow them. The metal cylinder protected the meat from grinding action of the stomach, yet when the hawks regurgitated the metal cylinders, the meat inside was found digested. This experiment conclusively refuted the earlier theory of digestion by means of grinding action of the stomach. Which brilliant scientist conducted this experiment?

(a) Leonardo da Vinci (b) René Antoine Ferchault de Réaumur (c) Van Helmont (d) Jan Ingenhousz

711. In 1952, when still a graduate student, he circulated water, ammonia, methane and hydrogen past an electric discharge to simulate the ultraviolet radiation of the sun and got complicated molecules associated with life such as glycine and alanine. This experiment was one of the first great experiments which proved that life indeed could arise from non-living matter. Who was this scientist?

(a) Carl Sagan (b) Murray Gell-Mann (c) Donald Arthur Glaser (d) Stanley Lloyd Miller

712. He transferred many colonies of bacteria to a medium containing penicillin. Only a few resistant strains could grow supporting the Darwinian view that these bacteria contained some mutant genes which conferred on them the ability to survive the action of penicillin. Who was this scientist?

(a) Joshua Lederberg (b) Daniele Bovet (c) Severo Ochoa (d) Arthur Kornberg

713. In 1929, when he was just 25 years old, this young German surgeon introduced a hollow tube in his

arm vein and pushed its tip up to the heart. Thus he showed to the medical world a unique way of putting drugs directly into the heart. Who was this pioneer experimenter?
(a) André (b) Frédéric Cournand Werner Forssmann (c) Dickinson Richards Jr. (d) George Wells Beadle

714. In the beginning, the above-named experimenter could not find a use for his new technique, so he approached a world famous thoracic surgeon to ask if he coud find a use for his new technique. He replied with this famous remark, 'I run a clinic for patients, not a circus'! Who was this thoracic surgeon?
(a) Hermann Joseph Muller (b) Ernst Boris Chain (c) Ferdinand Sauerbruch (d) Joseph Erlanger

715. This is a sculpture made by Monteverde for the Paris Exhibition, 1878, and represents one of the greatest experiments done in the human history. What does it depict?

716. Which biologist first showed that if a cock's testicles were excised and transplanted into its abdominal cavity, it retained its secondary sexual characteristics including a normally developed comb?
    (a) Arnold Adolph Berthold (b) Robert Debré (c) Charles Loomis Dana (d) John Dalton

717. Which famous biologist attenuated the germs of anthrax disease by heating them and inoculated sheep with them, thus making the sheep resistant to anthrax?
    (a) Joseph Lister (b) Louis Pasteur (c) Lazaro Spallanzani (d) Francesco Redi

13
GREAT BIOLOGISTS

718. He was accomplished in music, art and six languages when he left home for law school in 1765. He gained reputation as a writer but was also a great biologist. He founded the science of morphology and his work on plants foreshadowed Darwin's. Name him.
(a) Johann Wolfgang Von Goethe (b) Percy Bysshe Shelley (c) Gottfried Wilhelm Leibniz (d) George Wilhelm Friedrich Hegel

719. Charaka is usually known as the father of Indian medicine. However it is not generally known that he also tried to list the names of all animals and plants then known to science. How many animals did he list in his *Charaka Samhita?*
(a) 200 (b) 750 (c) 1,000 (d) 2,002

720. How many plants did Charaka list in *Charaka Samhita?*
(a) 50 (b) 101 (c) 340 (d) 501

721. Who developed the current scientific system of naming species with two names — the binomial system of nomenclature?
(a) Karl Ernst Von Baer (b) Robert Hooke (c) Anton Van Leeuwenhoek (d) Carolus Linnaeus

722. Who was the first to describe the optic nerve and the Eustachian tube?
(a) Alcmaeon of Croton (b) Hippocrates (c) Aristotle (d) Galen

723. This great Renaissance scientist showed the double curve of the spine and facial sinuses for the first time and derived advanced theories of physiology. But

fearing staunch opposition, he did not publish his findings, keeping them hidden in coded notebooks. Who was he?
(a) Andreas Vesalius (b) Leonardo da Vinci (c) Bartolommeo Eustachius (d) Hans Karl Von Euler-Chelpin

724. He was physician to Henry II of France and Catherine de Medici. He was the first modern scientist to take up dissection as an important part of a physician's training. Who was he?
(a) Nicolo Massa (b) Hieronymus Bock (c) Thomas Vicary (d) Jean Fernel

725. Which scientist first extracted 'rennet' from a calf's stomach, helping to make cheese more easily?
(a) Richard Kuhn (b) Christian Hansen (c) Karl August Folkers (d) George Wald

726. Before William Harvey discovered circulation, the same conclusions had been derived by another great biologist, but so contradictory were they to popular belief that he refused to accept them. Who was he?
(a) Hieronymus Fabricius (b) Santorio Sanctorius (c) René Descartes (d) Jan Baptista Van Helmont

727. Who was the first biologist to apply mathematics to a biological problem?
(a) René Descartes (b) William Harvey (c) Issac Newton (d) Aristotle

728. Which scientist saw capillaries for the first time under a miscroscope?
(a) Anton Van Leeuwenhoek (b) Hieronymus Fabricius (c) Marcello Malpighi (d) Jan Swammerdam

729. In which structure did the above scientist see the capillaries?

(a) Under human nails (his own) (b) In the lung tissues of a frog (c) In a rabbit's retina (d) In a guinea pig's skin

730. Who discovered the red blood corpuscles?
(a) Jan Swammerdam (b) Olof Rudbeck (c) Georges Louis Buffon (d) Carolus Linnaeus

731. In which year did the abovenamed scientist announce his discovery of red blood corpuscles?
(a) 1601 (b) 1631 (c) 1658 (d) 1699

732. Who discovered tiny ovarian follicles in animal ovaries?
(a) Erasistratus (b) Regnier de Graaf (c) James Lind (d) Jesse William Lazear

733. Who discovered the cell nucleus?
(a) William Sturgeon (b) Anton Van Leeuwenhoek (c) Nicolas Le Blanc (d) Robert Brown

734. In which structure did the abovenamed scientist see the cell nucleus?
(a) Cells of the orchid root (b) Pollen grains (c) Red blood cells of humans (d) White blood cells of humans

735. His most significant work was on the process of digestion. Among the many ingenious experiments he conducted, was the one in which he allowed a hawk to swallow a sponge and after regurgitation, squeezed the juice out. This fluid could dissolve meat. Who was this great biologist?
(a) Antoine Laurent Lavoisier (b) René Antoine Ferchault de Réaumur (c) Giovanni Battista Morgagni (d) Hermann Boerhaave

736. Which biologist discovered the intracellular apparatus which appears to be concerned with secretion and with the linking of carbohydrate groups to protein molecules in the formation of glycoproteins?

(a) Harry Goldblatt (b) Oliver Goldsmith (c) Camillo Golgi (d) Philemon Holland

737. Which famous biologist showed in the middle of the nineteenth century that beer and butter milk are products of fermentation brought about by yeast?

    (a) Louis Pasteur (b) Friedrich Henle (c) Ignaz Philipp Semmelweiss (d) Joseph Lister

738. Which famous biologist coined the phrase 'the origin of life' and made it respectable?

    (a) Wendell Meredith Stanley (b) Svante August Arrhenius (c) Stanley Lloyd Miller (d) Alexander Ivanovich Oparin

739. Which of the following biologists has been dubbed as the 'architect of molecular biology'?

    (a) Jacques Lucien Monod (b) Frederick Sanger (c) Francis Crick (d) James Watson

740. Who is generally considered the founder of the science of comparative anatomy?

    (a) Charles Darwin (b) Jean Bapatiste Lamarck (c) Georges Cuvier (d) William Smith

741. He seemed a backward child in school and was made an apprentice to a barber and a shoemaker. Ultimately he got a chance to study medicine and reached the pinnacle of his career when he shared the 1906 Nobel Prize in Medicine and Physiology with Camillo Golgi. Who was he?

    (a) Wilhelm Von Waldeyer (b) Ramón y Cajal (c) Sir Ronald Ross (d) Charles Laveran

742. Which biologist first noticed that the muscles of dissected frog legs twitched wildly when a spark from an electric machine struck them?

    (a) Kaspar Friedrich Wolf (b) Franz Joseph Gall (c) Georges Cuvier (d) Luigi Galvani

743. Which of the following biologists can rightly be called the founder of modern embryology?
(a) Kaspar Friedrich Wolff (b) John Kidd (c) Andreas Vesalius (d) William Harvey

744. Which of the following biologists is known as the German Pliny?
(a) Libavius (b) Konrad Von Gesner (c) Leonhard Fuchs (d) Johann Bayer

745. His views were quite mystical and obscure. One of his irrational beliefs was that the skull of vertebrates is formed from the fusion of several vertebrae and jaw by the fusion of limbs attached to them. Identify him.
(a) Lorenz Oken (b) Pliny the Elder (c) Maurice Wilkins (d) Alfred Russel Wallace

746. Which biologist was the first to show that cells produced even after several divisions of the fertilized egg can, when carefully isolated, develop into completely normal embryos?
(a) V.M. Ingram (b) Andrew Fielding Huxley (c) Hans Driesch (d) Severo Ochoa

747. Who was the first scientist to synthesize a useful drug for syphilis, thus starting modern chemotherapy?
(a) Alexander Fleming (b) Paul Ehrlich (c) Emil Fischer (d) Charles H. Best

748. Who is usually considered the founder of botany?
(a) Theophrastus (b) Pytheas (c) Aristotle (d) Seneca

749. Who is usually considered the founder of zoology?
(a) Plato (b) Aristotle (c) Herophilus (d) Erasistratus

750. Which biologist was the first to study cilia and show that unicellular creatures could use these for locomotion?

(a) Rudolf Kolliker (b) Karl Gegenbaur (c) Karl Theodor Ernst Von Siebold (d) Matthias Jakob Schleiden

751. He became interested in childbed fever in the early stages of his career. To prevent this fever, he forced the doctors working under him to wash their hands in chloride of lime before attending to patients. Who was this physician?
(a) Ignaz Philipp Semmelweiss (b) Friedrich Henle (c) Karl Leuckart (d) Joseph Lister

752. Napoleon had a medal struck in this scientist's honour in 1804. Name this scientist.
(a) Franz Anton Mesmer (b) Luigi Galvani (c) Kaspar Friedrich Wolff (d) Edward Jenner

753. Which of the following scientists is generally regarded as the founder of histology?
(a) Ambroise Paré (b) Marie François Xavier Bichat (c) Joseph Lister (d) V.M. Inqram

754. This great biologist made many discoveries by sheer chance. When somebody made a tongue-in-cheek reference to this, he replied with this famous remark, "Chance favours the prepared mind." Who is this biologist?
(a) Charles Darwin (b) Aristotle (c) Louis Pasteur (d) Joseph Lister

755. Which surgeon dropped rocks and other heavy objects on the faces of cadavers and studied the fractures thus produced?
(a) René Le Fort (b) Ambroise Paré (c) Galen (d) Avicenna

756. He is sometimes called the father of Australia. Botany Bay in Australia is named, because this was the first point where he landed in search of botanical material. Who was this botanist?

(a) Mathew Flinders (b) Robert Brown (c) Sir Joseph Banks (d) Charles Darwin

757- Identify these scientists and their discoveries.
763.

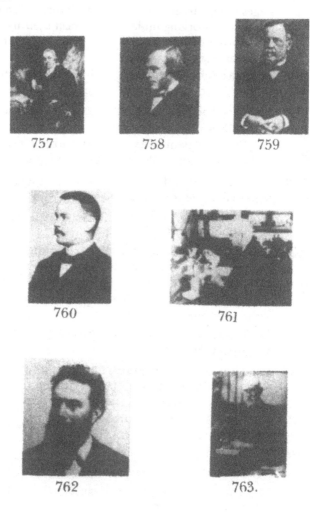

764. Which biologist was the first to show that green plants take up carbon dioxide and give off oxygen in the presence of light?
(a) J.E. Purkinje (b) Jan Ingenhousz (c) Stephen Hales (d) Hermann Helmholtz

765. Which of the following scientists is best known for his work on photosynthesis?
(a) Melvin Calvin (b) Erwin Mueller (c) Peter Debye (d) Leonardo Da Vinci

766. Which famous biologist propounded the 'germ theory of disease'?
(a) Robert Koch (b) Thomas Syndenham (c) Joseph Lister (d) Louis Pasteur

767. This great scientist was once travelling in a carriage with snow all around. Suddenly an idea occurred to him that snow could delay putrefaction of dead bodies. He jumped out of the carriage, bought a chicken and stuffed it with snow to test his idea. But in the process, he caught a chill and died later. Who was this brilliant and impetuous scientist?
(a) Roger Bacon (b) Francis Bacon (c) Aristotle (d) Anton Van Leeuwenhoek.

14
BIOLOGY IN EVERYDAY LIFE

Agriculture

768. When did agriculture originate?
 (a) 2,000 years ago (b) 7,000-13,000 years ago
 (c) 20,000-27,000 years ago (d) More than 50,000 years ago

769. This crop was first cultivated in the Peruvian and Bolivian Andes in about A.D. 200, and was introduced to Europe by the Spaniards in the late 16th century. Name it.
 (a) Oats (b) Rice (c) Potatoes (d) Rye

770. Any organism that damages the economic and physical well-being of humans is called a
 (a) parasite (b) commensal (c) exoparasite (d) pest

771. Chemicals that kill insects, weeds, disease organisms, ticks, mites and rats are collectively known as
 (a) herbicides (b) pesticides (c) insecticides
 (d) harmful chemicals

772. To which class of pesticides do DDT, BHC, aldrin and endosulphan belong?
 (a) Organochlorines (b) Organophosphates
 (c) Carbamates (d) Biopesticides

773. Which pesticide represents about 50 per cent of the total volume of pesticides used in India?
 (a) DDT (b) Baygon (c) Tik-20 (d) BHC

774. Several artificial methods of vegetative propagation are practised by horticulturists these days. In which technique are the roots induced on a stem before it is detached from the parent plant for propagation?

(a) Cutting (b) Grafting (c) Layering (d) Rooting

775. In one technique of artificial propagation of plants, a ring of bark tissue is removed and covered with moist moss or cotton and wrapped with a polythene sheet. After the injured part produces roots, the branch is cut and planted separately. What is this technique known as?
(a) Mound layering (b) Air layering (c) Cut layering (d) Graft layering

776. Grafting is the art of joining parts of plants such that they grow as one plant. What is the name given to the part of the graft which gives rise to the upper portion?
(a) Scion (b) Stock (c) Bud (d) Meristem

777. What is the name given to the pale yellow resin obtained from the bark of *Pistacia lentiscus* and which is used for making varnish?
(a) Mangosteen (b) Lac (c) Sadge (d) Mastic

778. Which of the following grains has been given the name 'pampered corn'?
(a) Wheat (b) Jowar (c) Ragi (d) Maize

779. In plant breeding, what does the term 'emasculation' mean?
(a) Crushing the pollens of a flower (b) Removing the sepals (c) Removing the stamens (d) Removing the petals

780. From what kind of tree is turpentine obtained?
(a) Most kinds of pine tree (b) From most trees (c) From grasses only (d) From oil wells

Animals in the service of man

781. Which is generally considered to be the first animal domesticated by man?
(a) Hen (b) Horse (c) Buffalo (d) Dog

782. Who or what are the Aurochs?
(a) The ancestors of European cattle (b) Places where cattle are kept hygienically (c) Special breeds of dogs used for hunting (d) Breeds of hens which lay more than a dozen eggs in a single day

783. Breeding and management of insects for the production of silk is known as
(a) mymiology (b) sericulture (c) apiculture (d) pisciculture

784. What is apiculture?
(a) Artificial breeding of apes to save them from extinction (b) Growing grapes as a cash crop (c) Bee-keeping (d) Artificial breeding of lac insects

785. What is pisciculture?
(a) Bringing up lions in a zoo (b) Activities undertaken to save pandas from extinction (c) Putting cattle to farm work (d) Production of fishes

786. Production of useful aquatic plants and animals such as shrimps, prawns, fishes, lobsters, snails and crabs, etc., is known as
(a) hydraulic agriculture (b) hydrophiliculture (c) aquaculture (d) agropisciculture

787. Around what time did man domesticate the pig, horse and donkey?
(a) 6000 B.C. (b) 3000 B.C. (c) 100 B.C. (d) A.D. 100

788. What does the caterpillar of *bombyx mori* produce?
(a) Wool (b) Deposits of sulphur (c) Silk (d) Cashmere

Biotechnology

789. Organisms which can bring about soil nutrient enrichment are known as
(a) soil bacteria (b) biofertilizers (c) sympathetic bacteria (d) proxy fertilizers

790. The proper definition of biotechnology is
 (a) modern technology used in biological laboratories (b) technology used to improve species of animals (c) use of living organisms and of substances obtained from them in industry (d) advanced technology used in biological warfare

791. Which is generally regarded as the first product of ancient biotechnology?
 (a) Alcohol (b) Opium (c) Rubber (d) Wax

792. How was cheese produced during ancient times?
 (a) By using the stomach of sheep and goats (b) By using sap of fig trees (c) By boiling the milk along with skin of rabbit (d) By letting the milk stay outside the house overnight with a cat hair inside it

793. What is this structure?

794. Antibodies made outside the body by hybrid cell cultures are known as
 (a) hybrid antibodies (b) specific antibodies (c) biotechnological antibodies (d) monoclonal antibodies

795. Who first thought of producing antibodies by hybrid cell cultures or hybridomas?
(a) Hargobind Khorana (b) César Milstein and Georges Köhler (c) Michael Brown and Joseph Goldstein (d) Susumu Tonegawa

Medical applications

796. The world's first successful human heart transplant operation was performed by Dr. Christian Barnard. Name the individual whose heart was used?
(a) Miss Denise Ann Darvall (b) Louis Washkansky (c) Doe Ryong (d) Jim Reeves

797. Who invented Hover-bed for severely burnt patients, which was essentially a Hovercraft turned upside down, so that it blew air upwards in order to support a patient on air?
(a) Robert Lee Wild (b) Dr. Richard K. Wampler (c) Dr. J.T. Scales (d) Willem Kolff

798. Fluosol DA is a derivative of petroleum and milky white in colour. What is it called in common parlance?
(a) Artificial semen thickener for sterile couples (b) Artificial bile for digestion (c) A new and effective laxative (d) Artificial blood

799. Which of the following is an artificial heart?
(a) Jarvik 7 (b) Jarvik 8 (c) Pen State (d) Bue-cherl system

800. Who was the first patient to receive an artificial heart?
(a) Charley Wild (b) Barney Clarke (c) Thomas Willing (d) Louis Jones

801. Doctors need to look inside the human body for diagnosing illnesses. Which of the following imaging systems was developed in 1973 by Paul C. Lauterbur of the State University of New York at Stony Brook?

(a) Positron emission tomography (b) Computerized axial tomography (c) Digital x-ray imaging (d) Nuclear magnetic resonance

802. Which revolutionary imaging system was developed by Godfrey Hounsfield and Allan Cormack?
(a) Computerized axial tomography (b) Digital x-ray imaging (c) Position emission tomography (d) Digital subtraction angiography

803. The following pictures show a new and revolutionary medical technique. Name it.

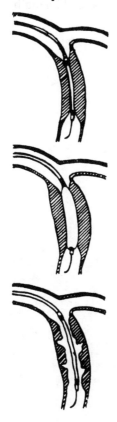

804. Dr Jonas Salk and Dr Albert Sabin, who actually developed vaccines for poliomyelitis, never got the Nobel Prize. Instead the Nobel Prize regarding control of poliomyelitis was awarded to another scientist about whom Dr. Salk remarked, 'He pitched a very long forward pass and I happened to be in the right spot to receive it.' Name this scientist.
(a) John Franklin Enders (b) Axel Hugo Teodor Theorell (c) Sir Hans Adolf Krebs (d) Fritz Albert Lipmann

805. What are these strange looking bands?

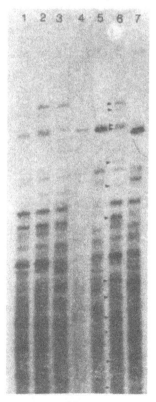

806. This is a photograph of historic significance. Can you identify it?

807. Name the condition of the eye which causes round objects to appear as though they were oval?
(a) Myopia (b) Presbyopia (c) Astigmatism (d) Retinitis

808. Who invented the cardiac pacemaker?
(a) Ake Senning (b) Robert Mondd (c) J. Leonard Corning (d) Carl Koller

809. Iron lung is an artificial device to assist respiration in an unconscious patient. Who invented it?
(a) Spembly Llord (b) Philip Drinken (c) Joseph Lister (d) Ignaz Phillip Semmelweis

810. Who invented electrocardiography (ECG), a device which helps diagnose heart's ailments?
(a) Niels Ryberg Finsen (b) Herbert Spencer Gasser (c) Joseph Erlanger (d) Willem Einthoven

811. Who was the first scientist to record the electrical activity of the brain?
(a) Charles Guillaume (b) Nicola Tesla (c) Hans Berger (d) Michael Faraday

Bioenergy

812. The total amount of living material in a unit of area is known as
(a) bioenergy (b) biomass (c) biovegetation (d) biolife

813. What per cent of solar energy reaching the earth's surface is converted into biomass?
(a) 100% (b) 50% (c) 2% (d) 0.2%

814. Coal, petroleum and natural gas are examples of
(a) Fossil fuels (b) biofuels (c) biogas and bioproducts (d) animal energy

815. One of the forms in which animal energy is available is human muscle power (HMP). How much HMP is used in India as compared to the total electricity generated per year in India?
(a) 1/2 (b) 1/5 (c) 1/10 (d) 1/20

816. What is draught animal power or DAP?
(a) Total calories required to feed all animals in India (b) Electricity generated by animals (c) Average power of each Indian animal (d) Total power which can be generated by animals in our country

817. Which of the following woods is considered better as a fuel?
(a) Gymnospermous (b) Monocotyledonous (c) Dicotyledonous (d) Wood from tall pteridophytes

15
GREAT BOOKS

818. Who wrote the epoch-making book on human anatomy, *De Corporis Humani Fabrica* (On the Structure of the Human Body)?
    (a) Nicolaus Copernicus (b) Andreas Vesalius (c) Francesco Redi (d) Karl Wilhelm Von Nageli

819. In which year was the above book written?
    (a) 1501 (b) 1526 (c) 1543 (d) 1598

820. In the above book, the drawings of the human anatomy are so beautiful and accurate that they are still reproduced today. Who made those drawings?
    (a) The author himself (b) Titian (c) Botticelli (d) Jan Stevenzoon Van Calcar

821. Who wrote the book, *De Motus Cordis* (Concerning the Motion of the Heart)?
    (a) William Harvey (b) Leonardo da Vinci (c) Jan Swammerdam (d) Galen

822. In which year was the above book published?
    (a) 1543 (b) 1576 (c) 1628 (d) 1699

823. Who wrote the book *Micrographia*, which contains some of the most beautiful drawings of microscopic observations ever made?
    (a) Sir Christopher Wren (b) Robert Hooke (c) Anton Van Leeuwenhoek (d) Ivan Pavlov

824. In which year was *Micrographia* written?
    (a) 1631 (b) 1665 (c) 1691 (d) 1699

825. Which Russian biochemist wrote the book, *The Origin of Life*, which for the first time dealt with the problem of life's origin completely from a materialistic point of view?

(a) Andrei Sakharov (b) Nikolaj Nikolajevic Semenov (c) Peter Leonidovitch Kapitsa (d) Aleksandr Ivanovich Oparin

826. Who wrote the gigantic seven-volume work entitled *Natural History of Invertebrates* which founded modern invertebrate zoology?
(a) Jean Baptiste Lamarck (b) Charles Darwin
(c) Karl Theodor Ernst Von Siebold (d) Theodor Schwann

827. Who wrote *The Causes of Evolution*, which is considered a landmark in evolutionary theory?
(a) John Gregor Mendel (b) Jean Baptiste Lamarck
(c) J.B.S. Haldane (d) Charles Darwin

828. Who wrote the monumental *Principles of Geology*, which popularized the view that earth was older than 6,000 years old, thus laying down the foundation of the theory of evolution?
(a) Sir Charles Lyell (b) Sir Roderick Murchison
(c) Adam Sedgwick (d) Joseph Henry

829. This book published in 1859 is still regarded as one of the classics of science. Only 1,250 copies were printed and were snapped up on the first day of the publication. Name the book.
(a) *The Descent of Man* (b) *The Origin of Species*
(c) *Essay on Population* (d) *Life from Cell to Cell*

16
LIVING BEINGS AND THEIR ENVIRONMENT

830. All plants, animals and other organisms that make up a distinct natural community in any climatic region is known as the
 (a) biomagnification (b) biounit (c) biome (d) biological ecosystem

831. Living and non-living elements of an environment that function together as a system is known as
 (a) ecosystem (b) biosystem (c) bioecosystem (d) ecoline

832. Which of the following terms best describes lakes or ponds that are rich in nutrients like phosphates and nitrates and are thus able to support a dense population of plankton?
 (a) Dystrophic (b) Eutrophic (c) Heterotrophic (d) Springs

17
CURIOUS FACTS

833. He was primarily a botanist, but while looking at a suspension of pollen in water made an epoch-making discovery in physics. Who was he?
(a) Robert Brown (b) Jean Baptiste Biot (c) William Nicol (d) André Marie Ampère

834. On 16 June 1822, he was accidentally wounded by a discharge from a musket causing a small opening in the stomach which could be manipulated from outside. For next sixty years an American military surgeon conducted experiments from this small opening and added substantially to the knowledge of physiology of digestion. Who was the patient?
(a) Frank Capra (b) Alexis St. Martin (c) Frederick Barber (d) John Singer

835. Which doctor conducted studies on the physiology of digestion in the above case?
(a) Ambroise Paré (b) Elmer Vernon McCollum (c) Konrad Lorenz (d) William Beaumont

836. To prevent people from plucking grapes, this farmer in Bordeaux, France, mixed some lime and copper sulphate with water and splashed it over his vines. His idea was to make the berries unappetising. Which major discovery did this simple act lead to?
(a) An efficient fertilizer (b) A nitrogen fixing agent (c) A fungicide (d) A rodenticide

837. When the large flightless bird Dodo became extinct on the island of Mauritius at the end of 18th century, a tree also lost its ability to grow. Which was this tree?

(a) Tambalacoque tree (b) Baobab tree (c) Raphia palm (d) Double coconut palm

838. What is special about the South American tree *brosimum utile*?

    (a) This is the only tree in the world which catches small mammals like rabbits (b) This tree is the only one to cry (c) It gives milk just like cow's milk (d) Its roots concentrate gold from gold salts in the earth and thus natives can sometimes find small chunks of gold within its roots.

839. During 1630s, Holland was gripped with mania for a particular flower. Which one?

    (a) Rose (b) Jasmine (c) Tulips (d) Orchids

840. The famous Crystal Palace, built in Hyde Park, London, to house the Great Exhibition of 1851 was inspired by a plant. Which one?

    (a) Rafflesia (b) Water lily (c) Oak (d) Pineapple

841. Which rose variety became the emblem of the Lancastrians during the 15th century Wars of the Roses in Britain?

    (a) Red Provins rose (b) Dog rose (c) Musk rose (d) Phoenician rose

842. Which tree derives its name from a common legend that Judas Iscariot hanged himself on it?

    (a) *Curchorus capsularis* (b) *Araucaria araucana* (c) *Quercus robur* (d) *Cercis siliquastrum*

843. Which tree has branches having the appearance of roots, giving rise to an ancient belief that the devil had turned the tree upside down?

    (a) Bristlecone pine (b) Cactus (c) Baobab (d) Elm

844. The popular name of which tree is derived from a remark by an anonymous joker that 'it would puzzle a monkey to climb that tree'?

    (a) Mahogany (b) Chile pine (c) Medlar (d) Poplar

845. Which of the following plants uses a battery of natural airguns to disperse its dust-like spores?

    (a) Sphagnum moss (b) Clubmoss (c) Horsetail (d) Liverwort

846. The giant redwoods of California are the tallest plants reaching a height of 110m. But which are the longest plants in the world?

    (a) Sorrel (b) Ressurrection plant (c) Liane (d) Rattan palm

847. Phosphorescent Bay near Parguera in Puerto Rico is so named because of the glow from certain organisms. Which ones?

    (a) Fireflies (b) *Pyrodinium* (c) *Protea repens* (d) Bird's nest clubmoss

848. Which plant seeds inspired the invention of Velcro fasteners?

    (a) Burdock (b) Breadfuit (c) Brazil nut (d) Larch

849. Which scientist fought with distinction in the Seven Years War, and received an officer's commission for bravery?

    (a) Sir Joseph Banks (b) Kaspar Friedrich Wolff (c) Jean Baptiste Lamarck (d) Edward Jenner

850. Which 20th century Russian biologist upheld the already obsolete Lamarckian theory of evolution and fitted his arguments so nicely around Soviet economic and philosophic theories that his controversial views were officially endorsed by the Communist Party in 1948?

    (a) Nikolai Ivanovich Vavilov (b) Trofim Denisovitch Lysenko (c) Lev Davidovich Landau (d) Peter Kapitza

851. Which famous Russian biologist was condemned to death for spying for England in a mock trial in July 1941?

(a) Nikolai Ivanovich Vavilov (b) Theodosius Dobzhansky (c) Alexander Ivanovich Oparin (d) Aharon Katzir-Katchalsky

852. Which famous scientist remarked, 'Had my lab been as up-to-date as those I have visited, I would never have made this discovery.'?
(a) Charles Darwin (b) John Gregor Mendel (c) Alexander Fleming (d) Stanley Miller

853. Which famous biologist ended his life in disappointment when he was proved wrong?
(a) Francesco Redi (b) Erasmus Bartholin (c) William Gilbert (d) Luigi Galvani

854. Which of the following plants was used as a wound dressing as late as World War I?
(a) Hornworts (b) *Sphagnum* (c) Copper moss (d) Red algae

855. The Edinburgh physician, Sir James Simpson (1811-70), was the first to show that chloroform was safe and effective, yet chloroform did not catch on. What did Queen Victoria do to make it popular?
(a) She knighted Dr Simpson (b) She made the use of chloroform mandatory by law (c) She allowed the use of chloroform in two of her confinements (d) Anyone opting for surgery under chloroform was to be given a job at the court

856. Who discovered the parathyroid glands while dissecting a rhinoceros?
(a) Andreas Vesalius (b) Galen (c) Sir Richard Owen (d) Anders Olaf Retzius

857. Although it may appear odd, as late as 1925, a biology teacher in Dayton, Tennessee, was brought to trial in a court of law for regularly teaching Darwin's theory of evolution in his classrooms. Can you identify this teacher?

(a) William Jennings Bryan (b) John T. Scopes (c) Isaac Asimov (d) Aldous Huxley

858. It is now possible to take two mouse embryos from different parents at an early stage of development, say, at the eight-cell stage, fuse them and develop them into a single mouse of normal size. What is this mouse called?

(a) Heterogeneous mouse (b) Allotype (c) Chimaera (d) Hybrid

859. This scientist is credited with the discovery of enzymes and was even awarded Nobel Prize in 1907. But he died during World War I, on the Rumanian front when he was asked to fight on the side of the Central Powers. Who was he?

(a) Eduard Buchner (b) Kotaro Honda (c) George Washington Carver (d) Walter Hermann Nernst

860. Which plant holds its leaves in a vertical position and aligns its edges north to south?

(a) Clver (b) Cinnamon (c) Cedar (d) Compass plant

861. The cork tissue is made up of flattened, thin-walled cells whose walls are permeated by a waxy substance known as

(a) suberin (b) elastin (c) collagen (d) reticulin

862. The seeds of which tree are so uniform in weight that they were once used as standard weights by goldsmiths?

(a) Oak (b) Banyan (c) Mulberry (d) Carob

863. According to a Greek legend, when infant Heracles was suckling at Hera's breast, he bit it inadvertently. Enraged, Hera pulled the infant away so violently that her milk gushed forth. Some drops of her milk turned into flowers. Identify this flower?

(a) White rose (b) Jasmine (c) White lily (d) Opium flower

864. Which famous biologist was trained as a lawyer, at which work he was so unhappy that he even attempted suicide?

(a) Theodor Schwann (b) Matthias Jakob Schleiden (c) Karl Nägeli (d) Karl Gegenbaur

865. A famous scientist was colour blind and was the first to describe colour-blindness. Colour-blindness is often named after him. Who is he?

(a) John Dalton (b) Galileo (c) Newton (d) Aristotle

18
THE HUMAN BODY

866. Regarding the need to understand the structure of the body, which famous scientist remarked, 'It is highly dishonourable for a reasonable soul to live in so divinely built a mansion as the body she resides in altogether unacquainted with the exquisite structure of it.'?
(a) Robert Boyle (b) Andreas Vesalius (c) Hippocrates (d) Aristotle

867. Where in the human body would you find the hamate bone?
(a) Skull (b) Ankle (c) Wrist (d) Chest

868. Where in the human body would you find ilium?
(a) Near the hip (b) In the abdomen (c) Leg (d) Chest

869. The connective tissue membrane that surrounds a bone is known as
(a) Mucosa (b) Epithelium (c) Mesothelium (d) Periosteum

870. The pair of veins which carry deoxygenated blood away from the head and neck are known as
(a) jugular veins (b) subclavian veins (c) hepatic veins (d) cerebral veins

871. Name the tube which connects the middle ear with the back of the throat
(a) Fallopian tube (b) Wolffian duct (c) Mullerian duct (d) Eustachian tube

872. The fluid that fills the space in front of the lens of the eye is known as

(a) tears (b) aqueous humour (c) vitreous humour (d) lachrymal fluid

873. The human brain is surrounded by cerebrospinal fluid which cushions it against external shocks. What is the name of the space which contains this fluid?
    (a) Subdural space (b) Subarachnoid space (c) Subpial space (d) Epidural space

874. The phase of the heart-beat cycle when the cardiac muscle contracts is known as
    (a) systole (b) diastole (c) protosystole (d) snap

875. What is the major substance making up the human body?
    (a) Proteins (b) Fats (c) Water (d) Carbohydrates

876. What are biochemical reactions in the body known as?
    (a) Catabolic reactions (b) Metabolic reactions (c) Anabolic reactions (d) Vital reactions

877. In which form is energy stored in the body?
    (a) ATP (b) Glucose (c) Heat (d) Sucrose

878. Why are carbohydrates essential to life?
    (a) They give structure to the body (b) They speed up chemical reactions (c) They provide energy (d) They keep the pH of the body balanced

879. Where would you find the metatarsal bones?
    (a) Hands (b) Skull (c) Chest (d) Foot

880. What is the normal range of human hearing in terms of frequency?
    (a) 100-1,000 Hz (b) 20-2,000 Hz (c) 20-20,000 Hz (d) 2,000-10,000 Hz

881. Which enzyme present in perspiration and tears destroys the cell walls of many bacteria, thus protecting our bodies against them?
    (a) Phosphatase (b) Lipase (c) Papain (d) Lysozyme

882. Compounds released by WBCs, which raise the body's temperature, are known as
(a) pyrogens (b) catecholamines (c) insulin (d) prostaglandins

883. B-cells and T-cells of the body are involved in
(a) new bone formation (b) blood formation (c) defence mechanism (d) suppression of pain

884. B-cells are involved in
(a) cell-mediated immunity (b) humoral immunity (c) producing enzymes (d) producing killer hormones

885. Salivary, tear, gastric and intestinal glands are examples of
(a) embryonic tissue (b) exocrine glands (c) endocrine glands (d) simple glands

886. Pituitary, thyroid, ovaries and adrenals are examples of
(a) embryonic tissue (b) exocrine glands (c) endocrine glands (d) simple glands

887. Which tissue acts as a strong inextensible attachment of skeletal muscles to bones?
(a) Ligaments (b) Tendons (c) Fibrils (d) Sinusoids

888. Haversian canals are found in
(a) brain (b) liver (c) lungs (d) bones

889. Which of the following changes would you find in the disease polycythemia?
(a) Increase in total RBCs (b) Decrease in total RBCs (c) Increase in total WBCs (d) Decrease in total WBCs

890. How much hemoglobin is there in 100 ml blood of an average healthy man?
(a) 15 mg (b) 5g (c) 15g (d) 50g

891. Thoracic duct is the major vessel of the

(a) arterial system (b) venous system (c) nervous system (d) lymphatic system

892. Of the various types of vertebrae in man, name the type which are 12 in number.
 (a) Cervical (b) Thoracic (c) Lumbar (d) Sarcal

893. Tympanic cavity is another name for
 (a) inner ear cavity (b) ear canal (c) middle ear (d) cavity of the eyeball

894. Where would you find bundle of His?
 (a) Stomach (b) Intestine (c) Ear (d) Heart

895- There are twelve pairs of cranial (brain) nerves in
906. the human body. Below are given their names and numbers. Match the nerve with its number.

895. I	(a)	Trochlear
896. II	(b)	Glossopharyngeal
897. III	(c)	Accessory
898. IV	(d)	Hypoglossal
899. V	(e)	Olfactory
900. VI	(f)	Trigeminal
901. VII	(g)	Vagus
902. VIII	(h)	Facial
903. IX	(i)	Auditory
904. X	(j)	Abducens
905. XI	(k)	Oculomotor
906. XII	(l)	Optic

19
MISCELLANEOUS

907. What was the most classic example presented as evidence for the existence of spontaneous generation?
    (a) Appearance of worms in human faeces
    (b) Appearance of earthworms after rainy season
    (c) Appearance of maggots on decaying meat
    (d) Appearance of tadpoles in ponds after rainy season

908. The evolutionary history of an organism is known as
    (a) phylogeny (b) ontogeny (c) phyloontogeny (d) phyllotaxis

909. Where is the Birbal Sahni Institute of Palaeobotany located?
    (a) New Delhi (b) Kanpur (c) Aurangabad (d) Lucknow

910. Mineral substances crystallising and developing into patterns resembling the outline of plants are known as
    (a) fossils (b) pseudofossils (c) pellets (d) ores

911. Organs like seal's flipper, bat's wing, horse's foot, cat's paw and human hand look superficially different but have a fundamentally similar plan and contain approximately the same number of bones. Such organs are called
    (a) homologous organs (b) analogous organs
    (c) similar organs (d) conjugate organs

912. Organs like wings of an insect and those of a bird which perform the same function but are not similar in structural details are called

(a) functionally similar organs (b) metalogous organs (c) analogous organs (d) phyllologous organs

913. Who wrote the book *Worlds in the Making*, picturing a universe in which life had always existed and migrated across space, continually colonizing new planets?
(a) H.C. Urey (b) Richard Lerner (c) Friedrich Mohs (d) Svante Arrhenius

914. Moulds, mushrooms and puff balls are examples of
(a) algae (b) fungi (c) viruses (d) protozoa

915. Which scientist replaced the old two-kingdom grouping of living organisms into a new and more modern five-kingdom grouping?
(a) Robert Harding Whittaker (b) Georgii Nicolaevich Flerov (c) Allen Gardner (d) William Francis Giauque

916. Which famous scientist named his house at Cambridge the Golden Helix in commemoration of the discovery of double helical structure of DNA?
(a) Erwin Chargaff (b) Maurice Wilkins (c) Francis Harry Compton Crick (d) James Dewey Watson

917. Which plant has been named after the founder of Singapore?
(a) Wolffia (b) Rugel's plantain (c) *Dionea muscipula* (d) Rafflesia

918. In ancient Greece, wreaths of laurel leaves and branches were placed upon the heads of heroes. Which tree did they come from?
(a) Cherry laurel (b) Japanese laurel (c) Spurge laurel (d) Bay tree

919. Which well-known tree is known in Britain as Wellingtonia?
(a) Giant sequoia (b) Oak (c) Elm (d) Poplar

920. This instrument is frequently used in biological investigations. Name it.

921. Which biologist is said to have remarked, "God has an inordinate fondness for beetles"?
(a) Charles Darwin (b) Carolus Linnaeus (c) Hugo de Vries (d) J.B.S. Haldane

922. Peat is used as a fuel in some countries. Remains of which plant give rise to peat?
(a) Angiosperms (b) Ferns (c) Moss (d) Gymnosperms

923. When Darwin published his book *The Origin of Species* in 1859, the scientific world was taken by storm and even scientists opposed and mocked at him. Which famous biologist, in this time of crisis, stood by Darwin's side, calling himself Darwin's bulldog?
(a) Richard Owen (b) Thomas Henry Huxley (c) Ernst Heinrich Haeckel (d) Alfred Russel Wallace

924. Which famous hypothesis was proposed in the early 1970s by British scientists Dr James Lovelock and Dr Sidney Epton, in collaboration with the American biologist, Dr Lynn Margulis?
(a) Neodarwinism (b) Neolamarckianism (c) Origin of life on earth from extraterrestrial civilizations (d) Gaia hypothesis

925. An organism that is haploid throughout its life except as a zygote is known as a
(a) diplont (b) haplont (c) fetus (d) suborganism

926. Indole acetic acid is a
(a) good fertilizer (b) pesticide (c) chemical found in the retina of nocturnal animals (d) naturally occurring auxin

927. A gland that produces a secretion that passes along a duct to an epithelial surface is known as a
(a) ductless gland (b) endocrine gland (c) exorine gland (d) secretory gland

928. Inulin is a
(a) hormone (b) enzyme (c) polysaccharide food reserve of some higher plants (d) larval form of an insect

929. Which essential element is found in haemoglobin, porphyrins and cytochromes?
(a) Copper (b) Manganese (c) Magnesium (d) Iron

930. An antigen that induces antibody production in members of the same species, but having different genetic constitutions is called
(a) isoantigen (b) polyantigen (c) parallel antigen (d) convergent antigen

931. Which of the following chemicals is converted in animals to Vitamin D by ultraviolet radiation?
(a) Cholesterol (b) Ergosterol (c) Erepsin (d) Ascorbic acid

932. What is the name of the instrument which is used for cutting thin sections of biological material for miscroscopic examination?
(a) Electrotome (b) Diacutter (c) Microtome (d) Ripper

933. What is an enzyme?
(a) A chemical compound that helps in metabolism (b) A substance broken down in metabolism (c) A form of yeast (d) A special chemical found within the nucleus

934. What do vitamins do?
(a) Block enzyme action (b) Help enzyme action (c) Speed up the breakdown of compounds by adding water molecules (d) Tone up the body muscles

935. What is the name given to the phenomenon when animals pass the hot summer months in a torpid condition, without moving about or eating anything?
    (a) Hibernation (b) Aestivation (c) Calcination (d) Coma

936. When was the first test tube baby born?
    (a) 3 August 1975 (b) 25 July 1978 (c) 2 January 1981 (d) 4 February 1982

937. What was the name of the first test tube baby?
    (a) Louise Brown (b) Joe Maggio (c) Anne Hathaway (d) Dick Jones

938. In which hospital was the first test tube baby born?
    (a) All India Institute of Medical Sciences, New Delhi (b) Japanese Institute of Gynaecology, Tokyo (c) Oldham Hospital (d) New Hampshire Hospital, New York

939. Which of the following doctors made the first test tube baby possible?
    (a) Patrick Steptoe (b) Robert Edwards (c) T.D. Dogra (d) Dr Indira Hinduja

940. When was the first 'frozen' baby born?
    (a) 4 February 1982 (b) 11 April 1984 (c) 17 August 1987 (d) 20 December 1989

941. What was the name of the first 'frozen' baby?
    (a) Joe Maggio (b) Anne Hathaway (c) Dick Jones (d) Zoe

942. In which hospital was the first 'frozen' baby born?
    (a) Edinburgh Royal Infirmary, Edinburgh (b) General Hospital, Moscow (c) London Hospital Medical College, London (d) Queen Victoria Hospital, Melbourne

943. Which of the following doctors made the first 'frozen' baby possible?

(a) Linda Mohr (b) Alan Frounson (c) Patrick Steptoe (d) Robert Edwards

944. Which company marketed the world's first commercial sanitary towels?
    (a) Johnson & Johnson (b) Kimberley-Clark (c) Johnson and Nicholson (d) Warner Brothers

945. When were the sanitary towel marketed?
    (a) 1901 (b) 1911 (c) 1921 (d) 1931

946. Under what name were the world's first sanitary towels marketed?
    (a) Cleenex (b) Nap Kin-girl (c) Hygen (d) Kotex

947. Who invented the tampon?
    (a) Carl Gustaf de Laval (b) Earl Hass (c) Professor Hiroyasu Funakubo (d) None of them

948. Who invented the contraceptive pill?
    (a) Gregory Pincus (b) John Rock (c) Linda Mohr (d) None of them

949. In which year was the pill invented?
    (a) 1948 (b) 1950 (c) 1954 (d) 1960

950. Which was the first contraceptive pill to be marketed?
    (a) Enovid 10 (b) Norlestrin (c) Ovulen (d) Ovral

951. Which of the following civilizations is known to have used vaginal tampons for contraception?
    (a) Indus valley civilization (b) Mayans (c) Aztecs (d) Pharoahs of Egypt

952. Who is said to have invented the technique of vaginal douche for the purpose of contraception?
    (a) Richard Von Krafft-Ebing (b) Charles Knowlton (c) Virginia Johnson (d) None of them

953. Which doctor was responsible for the first Indian test-tube baby?
    (a) Dr. Indira Hinduja (b) Dr. Sunil Khanna (c) Dr. Padmavati (d) Dr. Anthony Busuttil

954. What was the name of the first Indian test tube baby?
    (a) Vandana (b) Shishir (c) Mayur (d) Harsha

955. When was the first Indian test-tube baby born?
    (a) 6 August 1986 (b) 7 September 1987 (c) 18 January 1988 (d) 31 December 1989

956. What is the name given to the liquid secreted by the mammary glands immediately and for the first few days after parturition?
    (a) Protomilk (b) Casein (c) Rennin (d) Colostrum

957. An association between two organisms in which one benefits and the other remains unaffected either way is known as
    (a) mutualism (b) parasitism (c) commensalism (d) symbiosis

958. When tadpoles undergo metamorphosis into adult frogs, their tails are lost. Which cell organelle is responsible for this?
    (a) Golgi complex (b) Lysosomes (c) Ribosomes (d) Nucleus

959. An association of a number of different interrelated populations belonging to different species in a common environment which can successfully survive in nature is known as a
    (a) successful community (b) biotic community (c) food web (d) commercial group

960. Which common disease is also known as hydrophobia?
    (a) Typhoid (b) Dengue (c) Rabies (d) Influenza

961. Bomb calorimetre is used for
    (a) measuring the energy of bombs (b) finding the effect of bombs on caloric content of crops (c) storing wheat to protect from possible nuclear warfare (d) estimating the energy content of foodstuffs per unit weight

962. It has been shown that haemoglobin has a high affinity for oxygen at high pH values. At lower pH values (more acid conditions) it tends to release oxygen. This phenomenon is called
(a) Bohr shift (b) bolting (c) dissociation (d) metabolic acidosis

20
ONE-LINERS

Some statements are given below. Tick T if they are true and F if they are false.

963. Enzymes retain their catalytic action even when extracted from cells. T/F
964. Colour blindness is about 8 times more common in females. T/F
965. Amyloplasts of potato store starch. T/F
966. Rabies can be spread by the bites of foxes, wolves and jackals. T/F
967. The new techniques of SQUID and MET can be used to study the brain in health and disease. T/F
968. Purkinje fibres are found in the brain. T/F
969. Pyridoxine (Vitamin B_6) is soluble in water. T/F
970. Radius is a bone found in the skull. T/F
971. Organic matter older than 7,000 years can be very accurately dated by radiocarbon dating. T/F
972. The term actinomorphy is used to describe radial symmetry in plants. T/F
973. Sesamoid bones develop as an ossification within a tendon of vertebrates, especially mammals. T/F
974. Cro-Magnon man lived in Africa about one million years ago. T/F
975. Electron microscope is the best device to study living tissues. T/F
976. Scanning electron microscope magnifies the object much more than a transmission electron microscope. T/F

21
PHOTOQUIZ

977. This diagram depicts one of the most famous experiments ever conducted in the history of science. Who conducted this famous experiment?

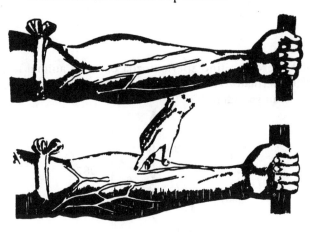

978. What are these strange looking objects?

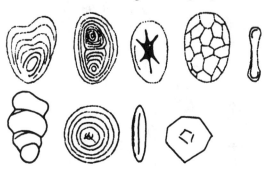

979. This is an enlarged view of a gland. Which one?

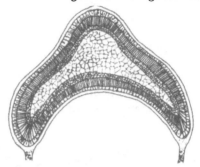

980- Here you find spermatozoa of some animal and
989. spermatozoids of some plants. How many of them can you identify?

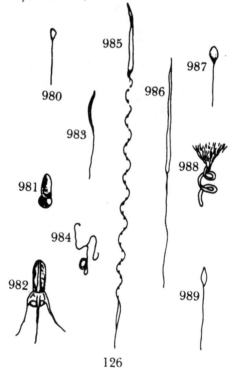

990. This series of four diagrams shows one of the most remarkable phenomena in the world of biology. Identify it.

22
PICTURE QUIZ

991. Identify this beautiful structure or creature.

992. Identify this photograph

993. What does this computer graphic show?

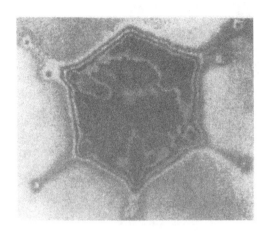

994. Identify this bird.

995. Identify this flower.

996. Identify this photograph.

997. Identify this photograph.

998. This photograph depicts one of the remarkable ways in which seeds are dispersed. Which seeds are these?

999. Identify this photograph.

1000. Identify this famous Indian biologist.

ANSWERS

1. STORY OF BIOLOGY

1. (c) Such greats as Aristotle and Isaac Newton, believed in it. The theory has been proved to be wrong now.
2. (a) From Greek *hepar* (liver), and *skopein* (to watch)
3. (b) The word artery in Greek means 'I carry air' These vessels were found to be empty in dead bodies.
4. (d) Alcmaeon, Greek physician flourished around 520 B.C.
5. (d)   6. (c)
7. He is Hippocrates (460 B.C.-377 B.C.), the Greek physician known as the father of medicine. He is the reputed author of over 87 medical treatises.
8. (d) According to some versions, Lamarck merely popularised the term.
9. (a)
10. (a) De'Luzzi (1275-1326) was an Italian anatomist.
11. (c) Quite surprisingly these 'homunculi', which of course do not exist, were seen, described and even drawn by microscopists with great enthusiasm and imagination!
12. (b) The biogenetic law stated that a developing embryo passes through stages similar to forms in the evolutionary history of its species: for example, a young human embryo has a fish-like form, with gill slits.

13. (d) Francis Bacon (1561-1626) was an En p 32 was scientist.

2. THE SCIENCES

14. (d)
15. (a) From Greek *morphe* (form) and *logos* (study of)
16. (b) Chemotaxonomy is widely used to study plant relationships.
17. (c) It differs from morphology in the sense that in the latter only external forms are studied.
18. (d) From Greek *entomon* (an insect)
19. (c) From Greek *oikos* (a house). Thus ecology literally means "study of organisms in their houses or environment."
20. (a) From Greek *palaios* (ancient)
21. (b)   22. (c)
23. (b) From Greek *dendron* (tree) and *chronos* (time)
24. (a)   25. (c)   26. (a)   27. (b)   28. (c)
29. (a) From Greek palynein (to strew or to scatter). Spores and pollens tend to get scattered easily.
30. (d)   31. (a)
32. (b) From Greet kytos (a hollow vessel). Cells were originally thought to be empty.
33. (d) From Greek *karyon* (a nut). Cell nucleus resembles a small nut.
34. (c) Derived from Latin *pomum* meaning fruit.
35. (a) From Greek *histos* meaning tissue.
36. (c) The term comes from Greek *emporos* (one who goes on shipboard as a passenger) and *iatrike* (medicine).
37. (α)This term was coined in 1948 by Norbert Wiener (1894-1964), an American mathematician.
38. (b) The word comes from Greek *phykos* meaning seaweed. The study is also known as algology.
39. (d) From Greek *ichthys* (a fish)
40. (c) From Greek *ethos* (character)

3. THE CELL

41. (d) Robert Hooke (1635-1703) was basically a physicist. He discovered the well known Hooke's law too.
42. (b)   43. (b)   44. (a)
45. (c) The word nucleus is derived from a Latin word *nux* meaning 'little nut'.
46. (a) and (b)
47. (c) Virchow epitomized his notion in a pithy Latin remark — *omnis cellula e cellula*, (all cells come from pre-existing cells)
48. (a) Such cells lack a nucleus as we commonly understand it. Bacterial cells belong to this class.
49. (c) These are more sophisticated cells.
50. (b)   51. (c)   52. (d)
53. (c) The word comes from the Greek *desmos* meaning chain or band.
54. (b)
55. (d) These were cells taken from the cancerous cervix of Henrietta Lacks. The name is a combination of the first two letters of her name. Although Mrs. Lacks died in 1951 at the age of 31, her cells are still multiplying in tissue culture and are used for research.
56. (c) Greek *mitos* means thread. Mitosis is so called because during this process chromatin is condensed into threads.
57. (a) The cells containing this substance are said to be 'lignified' and, since lignin forms an impermeable barrier, the cells are dead.
58. (b) Mitochondria contain their own DNA, RNA and even ribosomes, thus having a status of semiautonomous organelles. Mitochondrial DNA even directs the synthesis of some mitochondrial proteins. These considerations have led to the belief that during evolution, some-

how aerobic bacteria came to enter the cell and ultimately changed into mitochondria.

59. (d)
60. (c) These three are the typical changes seen in cell death. Pyknosis is shrinkage of nucleus, karyorrhexis is its fragmentation and karyolysis is its dissolution.
61. (b)
62. (a) M is mitosis and lasts for 1 hour; G_1 is the first gap lasting 8 hours; S is the synthesis phase during which new DNA is synthesized, and lasts for 8 hours; G_2 is the second gap lasting 4 hours.
63. (b) It is the interphase which is divided in G_1, S and G_2.
64. (a) This drug is obtained from *Colchicum autumnale*, the meadow saffron, and is used to treat certain cancers.
65. (d) Acid hydrolase can destroy virtually all the major components of the cell and hence it is enclosed in lysosomes.
66. Simple squamous. The cells are thin and plate-like. Example: lining of the lung.
67. Simple Cuboidal. The cells are cubical. Example: thyroid vesicles.
68. Simple Columnar. The cells are like columns. Example: gall bladder.
69. Pseudostratified ciliated columnar. Although cells comprise one layer, their nuclei are in different layers giving a false appearance of stratification. Example: trachea and large bronchi.
70. Transitional. Cells are arranged in layers. Deepest cells are column-like, superficial cells are more pear-shaped. Example: urinary bladder.
71. Stratified squamous non-keratinized. Cells in many layers. Top layer is flat. All cells are living. Example: vagina.

72. Stratified squamous keratinized. Cells as before but the top layers are dead forming a protective proteinaceous layer of keratin. Example: skin.
73. (c) Ordinary egg which we eat daily is a single cell although it is filled largely with the yolk.
74. (d) From Greek chondros (grain). So chondriosomes are literally "grain bodies".
75. (a)   76. (b)   77. (c)   78. (a)   79. (c)
80. (c)
81. (a) As a result of plasmogamy, two nuclei exist separately in the same cell, which is called a dikaryon (two nuclei).
82. (c) Stains used for electron microscopy contain heavy metals. Other useful stains are lead citrate and osmium tetroxide.
83. (b)
84. (a) and (c). Centriole is also found in motile sex cells of some primitive land plants.
85. (b) From Greek *diktyon* (net), and *soma* (body. Golgi complex often appears as a net like body.
86. (d) Lysosomes release hydrolases in damaged or ageing cells, killing them and hence the term.
87. (c)   88. (a)   89. (b)   90. (c)   91. (a)
92. (b)

CHROMOSOMES AND GENES

93. (b) Greek *chromos* means colour and *soma* means body. They take certain stains and show up as coloured bodies under the microscope. Chromosomes in their natural state are colourless.
94. (a) The karyotype offers a subtle tool in medical diagnosis.
95. (d) It is a plant and contains 1,262 chromosomes in each cell! In comparison man contains 46, dog 78, and *Amoeba proteus* 250.

96. (c) Common wheat with 42 chromosomes is a hexaploid. It means that there is a basic set of 7 chromosomes. Wheat cells contain six such sets. Adder's tongue fern and amoeba are also examples of polyploids.

97. (b) Also known as fruit flies. They breed quickly and prolifically and are less expensive to rear than other laboratory animals like rabbits or mice.

98. (a) Beadle was awarded Nobel Prize for his work in 1958.

99. (c) A gene pool helps the population for adapting itself to the environment.

100. (d)

101. (a) In contrast, inactive DNA is called heterochromatin.

102. (d) Euchromatin appears dark, while heterochromatin appears light. Q-banding is a method by which individual chromosomes can be identified.

103. (c)   104. (b)

105. (c) A good example of intermediate inheritance is sickle-cell trait, when a person having genes for both normal hemoglobin (A) and defective hemoglobin (S), produces both types of hemoglobin in about equal quantities. Normally only one gene — the dominant one — is expressed.

106. (a) A good example is Marfan syndrome when a single dominant gene produces long fingers and toes, flexible joints, disorders of the heart and of the eye.

107. (d) A good example is the gene for blue eyes which inhibits the expression of genes influencing the degree of melanin deposit in the iris.

108. (a) During an investigation into the effects of nerve stimulation, two scientists — Barr and

Bertram — noticed a small round mass of chromatin in the nerve cells of certain female cats. These clumps of sex chromatin are now aptly known as *Barr bodies*.

109. (c) In humans, satellites are present on chromosomes 13, 14, 15, 21 and 22.
110. (b) These groups are designated A to G. The chromosomes are arranged in order of decreasing length. Thus group A constitutes the longest chromosomes and group G the shortest.
111. (b) This chromosome was discovered in Philadelphia; hence the name. It is found in the majority of patients with chronic myelogenous leukemia and is thus a useful diagnostic tool.
112. (d) This is the latest kind of pollution being talked about.
113. (a) Because genes in a large population behave as beans in a bag. If black and white beans are kept in a bag and taken out at random, the results are quite similar to those of actual population genetics.
114. (b) The coding sequences are called exons.
115. (c)   116. (a)
117. (d) The method was developed by Giessen biochemist Robert Feulgen (b. 1884).
118. (b)
119. (a) "nm" stands for nanometer. 1 nm = 10^{-9} meters
120. (a)
121. (d) There are five major histones, known as H1, H2A, H2B, H3 and H4.
122. (c)

5. THE PROTEINS

123. (b) Proteins are of key importance to life.

124. (a)
125. (d) As happens on boiling an egg. This seemed to put the minds of ancient philosophers topsy-turvy.
126. (c)
127. (b) It was prepared from the glandular lining of stomach.
128. (a) R represents the side chain. Different side chains give rise to different amino acids.
129. (c) However, only 20 occur in proteins.
130. (b) The other four essential amino acids are isoleucine, lysine, trytophan and threonine. These 8 amino acids are essential for survival.
131. (b)   132. (c)
133. (b) Of these, scientists have been able to identify fewer than 2%.
134. (d)
135. (a) The reagent is heated with the test material. Production of a brick-red precipitate indicates the presence of protein. This reagent was first prepared by a Paris chemist Auguste Millon; hence the name.
136. (c)   137. (c)
138. (d) Apoenzyme is joined with a non-protein group known as co-enzyme to form holoenzyme, also known simply as enzyme.
139. (b) Collagen is present in bone, skin and cartilage.
140. (a) Sanger (b.1918), after 8 years of hard work, worked out the amino acid sequence of insulin (a protein) in 1953. For this, he was awarded Nobel Prize in 1958.
141. (c) The American biochemist Vincent du Vigneaud (pronounced as dyoo-veenyoh) synthesized this molecule in 1954, for which he was awarded Nobel Prize in 1955. Oxytocin contains only 8 amino acids.

6. VITAMINS

142. (a) Funk (1884-1967), an American biochemist of Polish origin, coined this name in 1912, which means a vital amine.
143. (d) Cats can synthesize their own Vitamin C.
144. (d)   145. (g)
146. (a) Niacin is known as Vitamin B_3 in the USA. However, formerly Niacin was known as Vitamin B_5 in Europe. Some textbooks still refer to Niacin as Vitamin B_5, but that is not correct.
147. (i)   148. (b)   149. (e)   150. (c)   151. (f)
152. (j)   153. (h)
154. (a) Although termed Vitamin H, it is actually a component of Vitamin B Complex.
155. (e)
156. (a) Angular stomatitis means, cracking of lips and corners of the mouth.
157. (d)
158. (c) Peripheral neuritis means inflammation of peripheral nerves.
159. (b)   160. (c)
161. (a) Also known as antixerophthalmic vitamin.
162. (b)   163. (e)   164. (d)   165. (e)   166. (d)
167. (a)   168. (b)   169. (c)
170. (b) PP stands for pellagra preventing.
171. (a)   172. (d)   173. (a)

7. ORIGIN OF LIFE

174. (c) The term *Biogenesis* was coined in 1870 by the English biologist Thomas Huxley.
175. (a)
176. (b) It doesn't have any takers now.
177. (b)

178. (d) By 180 million years ago, Pangaea had split into two great land masses — the northern Laurasia and the southern Gondwana.

8. MICRO-ORGANISMS

Bacteria

179. (a) A single such bacteria is called a bacillus. The term comes from the Latin word 'baculus' meaning rod.
180. (c) A single such bacteria is called a coccus. The term comes from the Greek word 'kokkos' meaning a berry, which is what these bacteria look like.
181. (b)
182. (d) Archaebacteria means ancient bacteria. They seem to be early forms of life.
183. (b) Halophile means salt loving.
184. (a) Because they produce methane. Methanogens are actually responsible for the production of methane in biogas fermenters.
185. (b) The singular is flagellum. The word comes from the Latin word "flagellare" meaning 'to whip'.
186. (c) Gram (1853-1938), a Danish physician, used crystal violet to stain the bacteria. The method is still known as Gram's staining method.
187. (b)   188. (d)   189. (a)   190. (b)   191. (c)
192. (d) All plasmids consist of circular double-stranded DNA molecules of molecular weight 10^6-10^8 daltons.
193. (a)   194. (b)   195. (c)   196. (a)

Viruses

197. (b) Beijerinck (1851-1931) was a Dutch microbiologist, who discovered viruses in a tobacco plant suffering from mosaic disease.

198. (d) It was first thought that an infectious and poisonous fluid is responsible for viral diseases — hence the name.
199. (c) The term comes from Latin *capsa* meaning box.
200. (a)   201. (c)   202. (b)
203. (d) Felix d'Herelle (1873-1949) was a French microbiologist.
204. Poxvirus (size — 200 nm, disease — smallpox).
205. Herpesvirus (size — 100 nm, disease — Herpes simplex).
206. Orthomyxovirus (size — 100 nm, disease — influenza).
207. Coronavirus (size — 100 nm, disease — common cold).
208. Bacteriophage (size — variable, up to 800 nm, disease — none).
209. Togavirus (size — 50 nm, disease — several heterogeneous diseases associated with fever).
210. Adenovirus (size — 80 nm, disease — common cold).
211. Picornavirus (size — 25 nm, disease — polio).
212. (a) Examples of retroviruses are Rous sarcoma virus (RSV) and HIV (the virus causing AIDS).
213. (c) In April 1986, ten prostitutes having antibodies against AIDS virus were detected in Madras.

Other micro-organisms

214. (c) It is tempting to think of them as the 'missing link' between bacteria and viruses but they are not.
215. (a) Named after the American microbiologist Howard Taylor Ricketts (1871-1910) who discovered them. Ricketts himself died of typhus.

9. HEREDITY AND EVOLUTION

General

216. (b) From Latin *at* (above) and *avus* (grandfather). Thus the term literally means "above grandfather".
217. (a) His statement showed his belief in the theory of the 'continuity of germ plasm'.
218. (d) Lamarck deserves full credit for moving evolutionary theory into the forefront of biological thinking. His proposed mechanism of evolution was however wrong.
219. (b) In this book he advanced his theory of evolution stressing on the inheritance of acquired characters.
220. (c) Examples are lion and the lioness, man and woman, peacock and peahen.
221. (a)
222. (b) They do not interbreed in natural conditions because of differences in geographical distribution.
223. (c) *Cycads* and *Gingko,* two primitive gymnosperms, have ciliated sperms like the pteridophytes.
224. (d)
225. (c) An example is the quills of both African and American porcupines.
226. (b) 227. (a)
228. (d) Darwin was chosen only as a companion for the captain, Robert Fitzroy. His ticket was his ability to afford the trip!
229. This is the skull of *Australopithecus boisei*, one of the earliest recognizable humans. *A. boisei* lived on earth 1-1.9 million years ago.
230. (a) An example of convergence is the wings of birds and insects.
231. (c)

Human evolution

232. (b)
233. (c) Homo erectus lived on earth from 1.9 million years ago to 300,000 years ago — a total period of 1.6 million years.
234. (a) Huxley (1825-1895) was a doctor by profession.
235. (d) Both Disraeli and Gladstone, the two great British Prime Ministers, were strong anti-Darwinists.
236. (b) Not only this, by a curious coincidence, Wallace (1823-1913) sent his paper to Darwin for opinion. It is said that when Darwin received the manuscript he was thunderstruck.
237. (c) Discovered in Germany's Neander Valley (hence the name), only a few pieces of skeleton were found, including a skullcap and leg bones.
238. (a) Owen (1804-1892) was an English zoologist.
239. (d) *A. afarensis* lived between 3-4 million years ago.
240. (b) *Proconsul* is not a hominid, but an ape-like creature from whom earliest hominids descended.
241. (c) Johanson and his team also discovered the fossils of a group now dubbed the First Family. Lucy lived sometime 3-4 million years ago.
242. (a)

10. DIVERSITY OF ANIMAL LIFE

General

243. (d) About 10,000 new animal species are being discovered every year.
244. (a)
245. (b) Out of a total of 1.2 million living animal species, 0.9 million belong to the phylum arthropoda.
246. (b) and (c) In mammals the legs are positioned under the body rather than at the sides. Their teeth are divided into three types: incisors, ca-

nines and molars. This division is not seen in reptiles.
247. (a) and (c) Sea squirts and fish are invertebrate and vertebrate chordates respectively. They both share the phylum chordata with man.
248. (b) The mesoderm is a layer between the inner and outer walls (endoderm and ectoderm) which forms first in a developing embryo.
249. (d)
250. (a) A good example is eel which can live in both fresh and salt water.
251. (b) The word comes from Latin *saltus* (a leap).
252. (d)   253. (c)
254. (a) This type of symmetry is characteristic of most free-moving animals such as man, where one end constantly leads during movement.
255. (c) Bee humming bird of Cuba is only 2¼ inches long and weighs less than 2 gm.
256. (d) They are also the first to possess ear drums.
257. (b) They include such organisms as the freshwater hydra and jellyfish.
258. (c)
259. (d) For example, park fallow deer.
260. (a)   261. (c)
262. (a) Animals belonging to coelenterata and echinodermata display this kind of symmetry.

Prehistoric animals

263. (c) Compsognathus measured only 2 ft in length and weighed around 3 kg — little more than a large chicken.
264. (c)   265. (a)
266. (d) Literally meaning large lizard. This 20-ft-long carnivore was described by Dean William Buckland in 1824.
267. (c) Also known simply as Baluchitherium, it resembled a gigantic rhinoceros, without horn. Its

overall length was 35 ft and it weighed 20 tonnes (African elephant weighs only 6 tonnes!).

268. (b) Literally meaning shoulder lizard, it weighed an estimated 78 tonnes which is equal to the weight 13 African elephants!

269. (a) The dodo was a large heavy bird weighing about 50 lb. Because it moved slowly it was fairly easy to catch and though once common on the island of Mauritius, it became extinct in about A.D. 1680. Visiting sailors, who killed them in large numbers for food, contributed to their extinction. Nowadays 'as dead as a dodo' refers to anything totally extinct.

270. (b) Cuvier (1769-1832), a French anatomist, also described several spectacular fossils. For these discoveries he is also known as the founder of palaeontology.

271. This is the skeleton of the fossil bird *archaeopteryx* which shows features of both birds and reptiles.

Animals without backbones

272. (a) Swammerdam (1637-1680), a Dutch naturalist, produced excellent studies of insect microanatomy. Unfortunately he became insane later.

273. (c)

274. (b) Formerly known as coelenterata, this phylum includes sea anemones, jelly fishes, sea pens and hydra. The name cnidaria comes from Greek *knide* meaning nettle. Most of these animals can sting.

275. (d) Five teeth of sea urchins are arranged in such a way as to give the shape of a lantern. Each tooth can move separately so that the entire machinery works as a food shredder.

276. Fly (sponging).

277. Grasshopper (biting and chewing).
278. Butterfly (siphoning).
279. Mosquito (sucking and piercing).
280. Honey bee (chewing and lapping).
281. (d)
282. (c) The cells of sponges do not interact to form true tissues. The skeleton of one species is sold as the bath sponge.
283. (a)   284. (c)
285. (b) Because they have a proboscis, an extensile tube which can be shot out with force to catch prey.
286. (b) Greek *nem rtes* means sea mymph. Phylum Nemertinea includes ribbon worms, long ribbon like unsegmented marine worms, brightly coloured and sometimes growing up to 90 ft long.
287. (a),(b) and (c)
288. (a) 1 mm in diametre and 3 ft in length, they look like a length of horse-hair and tangle themselves in knots — hence the term Gordian worms.
289. (c) Each day's growth in the coral's calcareous deposit is marked by a ring. Rings of corals, known to have lived 300 million years ago, revealed this startling fact.
290. (b)   291. (c)
292. (d) It is in Australia. It is 2,000 km long and at places 450 m thick.
293. (b)   294. (d)   295. (a)   296. (b)
297. (c) It varies in thickness from a thin membrane in *hydra* to a thick gelatinous mass in jelly fish.
298. (a) The females of the tropical stick insect (*pharnacia serratipes*) have been measured up to 330 mm.

299. (b) Queen termites are known to live for up to 50 years.
300. (c) Male cicadas have special tymbal organs at the base of abdomen which vibrate from 120 to 480 times a second. They produce sounds detectable at more than a quarter of a mile away!

301. (u) 302. (r) 303. (o) 304. (a) 305. (s)
306. (h) 307. (b) 308. (t) 309. (v) 310. (p)
311. (i) 312. (c) 313. (q) 314. (k) 315. (d)
316. (m) 317. (e) 318. (j) 319. (g) 320. (l)
321. (f) 322. (n) 323. (c) 324. (a) 325. (b)
326. (n) 327. (l) 328. (p) 329. (m) 330. (a)
331. (r) 332. (o) 333. (b) 334. (q) 335. (f)
336. (c) 337. (g) 338. (i) 339. (e) 340. (h)
341. (j) 342. (d) 343. (k) 344. (b) 345. (j)
346. (g) 347. (e) 348. (a) 349. (i) 350. (h)
351. (c) 352. (b) 353. (d) 354. (f)

Fishes

355. (c) Devonian period lasted from 395 million years ago till 345 million years ago.
356. (c) Out of this only 2,300 are fresh water species.
357. (a) Bony fishes or teleost fishes comprise almost 90% of all known fish species, and because of their efficient mode of swimming and feeding, became the most successful class.
358. (a) While most sharks prey on other fish, basking sharks and whale sharks feed mainly on plankton.
359. (b)
360. This figure shows a lamprey attached by its sucking mouth to a living fish, sucking blood and flesh. Lampreys belong to the smallest class of fishes — the jawless fishes *(agnatha)*.
361. (c) Cartilaginous fishes have their mouth on the undersurface of the body. If it were used to sup-

ply water to gills, it would supply only sandy or muddy water.
362. (b)
363. (a) Sharks often feed on wounded or unhealthy fish, which swim clumsily, making a noise in the water as they do so.
364. (c) The first specimen known to science was caught in 1938 off the coast of East Africa.
365. (c)
366. (a) The word comes from Greek *pelagos* meaning sea.
367. (b) 368. (f) 369. (k) 370. (h) 371. (j)
372. (c) 373. (a) 374. (d) 375. (b) 376. (e)
377. (i) 378. (g)

Ambhibians
379. (c) It was also the first quadruped and is called *Ishthyostega*. Its remains have been discovered in Greenland.
380. (d) Of these, 300 are tailed (newts and salamanders) and 2,000 tail-less (frogs and toads).
381. (a) Measures 1 m in length and weighs 12 kg.
382. (c) From Greek *oura* (tail), and *delos* (visible).
383. (b) Very few tadpoles have been known to attack and eat other tadpoles, so count (c) also as the right choice.
384. (c) Axolotls normally remain in a sexually mature but nevertheless larval form throughout their lives. Metamorphosis can be stimulated in the laboratory by treatment with thyroid hormone.
385. (d) Caecilians are blind, limbless, burrowing animals.
386. (a) They share the same habitat as earthworms. However, they can be distinguished from earthworms by their method of locomotion which is much more like that of a snake.

387. (b)
388. (c) Also known as arrowpoison frogs. The frogs are placed by a hot fire until the toxins are exuded.
389. (c) Also known as Surinam toad.
390. (a) Spadefoots live in dry habitats and hide in their burrows during the day.
391. (d)

Reptiles
392. (c) They descended from a group of amphibians, now extinct, called labyrinthodonts.
393. (d)
394. (a) From Greek chelone (a tortoise).
395. (b) From Greek squamosus (covered with scales).
396. (b)   397. (d)
398. (a) and (c)
399. (c) The upper part of the shell is the carapace. Scutes are scale-like plates on the outside of a turtle's shell. The word plastron comes from French *plastron* (a breast plate).
400. (a)
401. (d) It builds a nest by scraping together clay and leaves. These nests may be as much as a foot in height.
402. (a) (b) and (c), Draco is also called the flying lizard and lives in Malaysia. Fringed gecko is also known as flying gecko or parachute gecko and is native of south-east Asia. Golden tree snake is one of the four species of flying snakes living in south-east Asia. These living specimens are strong proof that reptiles ultimately developed capacity of flying and gave rise to birds.
403. (b)
404. (c) Also called hamadryad, it can grow to a length of 18 ft.
405. (a)

Birds

406. (d) Palaeontologically this period lies in Jurassic era which lasted from 195 to 136 million years ago. This era is interesting also because it was during this period that the single land mass generally referred to as Pangaea began splitting into continents. The date of splitting is usually put as 180 million years.

407. (d) Compare it with total known animal species, which is 1,200,000. The birds are divided in 28 orders of equal status comprising 157 families. Out of these, perching birds (passerines) include 5,100 species or 60% of the total number.

408. (b) It is 8 ft tall and weighs about 125 kg. A healthy human adult is only 6 ft tall and weighs about 100 kg.

409. (a) It weighs around 15 kg which is very close to the upper size limit above which flight is impossible.

410. Parrot (for grasping).

411. Crow (for grasping).

412. Woodpecker (for gripping trees)

413. Jungle fowl (spurs for fighting — in males only)

414. Eagle (for seizing prey)

415. Ostrich (for running)

416. Coot (for paddling)

417. Mallard (for swimming)

418. Jacana (for walking on lily pads)

419. Ptarmigan (for running on snow)

420. (c) It lives in Cuba and weighs only 2 gm.

421. (b) If a humming bird' were as large as a human and consumed energy at its own rate, it would use 155,000 calories in a single day! Average adult man uses only 3,500.

422. (a) It lives in Inaccesible Island in South Atlantic and is only the size of a newly-hatched domestic chick.
423. (c) It lives in Asia and can attain air speeds up to 106 miles/hour. About 50% of the world's flying birds cannot exceed an air speed of 40 miles/hour.
424. (a) Melanins are responsible for yellows, browns and black. White is the result of reflection alone.
425. Parrot (For tearing fruit and seeds).
426. Crossbill (Crossed mandible tips to extract the seeds from fir cone scales).
427. Finch (For eating soft seeds).
428. Woodpecker (Strong chisel-shaped bill for boring nest-holes).
429. Eagle (For tearing prey).
430. Kiwi (The long slender bill is for probing into soil)
431. Avocet (For sifting water, to feed on small crustaceans).
432. Duck (For catching fish).
433. Spoonbill (For feeding on small aquatic animals).
434. Flamingo (Sieve like bill to separate small food items from water).
435. (a)
436. (a) and (c). The pygostyle is the characteristic tail bone of modern birds. They have hollow bones to reduce body weight.
437. All. Occipital condyle is a bony knob at the back of skull to which first vertebra is attached. This arrangement gives greater range of head movement. Uncinate process helps to strengthen the rib cage.

438. Weaver birds (nests woven from grasses and hang from twigs).
439. Cassin's malimbe (conical structure of vegetable fibres).
440. Tailor bird (two large leaves 'oversewn' with strands of grass).
441. Goldfinch. (Cup shaped nest to keep the eggs).
442. Humming bird (cuplike shapes made of fine vegetable fibres and webs).
443. Woodpecker (hole in a tree lined with wood chips).
444. Oven bird ('oven' of clay).
445. Malee (builds a mound and covers the eggs).
446. Plover (a shallow hollow on the ground).
447. (c) About 100-65 million years old. Ichthyornis literally means fish bird. It was similar to the modern tern and ate fish.
448. (d) 449. (a)
450. (c) From Latin *raptor* (a robber).
451. (c) From Greek *syringos* (a pipe). The syrinx is pipe-shaped.
452. (a) The word comes from Latin *remus* (oar).
453. (a) It forms a slot in front of the leading edge of the wing and this reduces turbulence.
454. (c) 455. (c) 456. (b)
457. (c) Elephant birds lived in Madagascar until relatively recently. They probably weighed up to 450 kg!
458. (g) 459. (l) 460. (p) 461. (u) 462. (j)
463. (d) 464. (a) 465. (b) 466. (x) 467. (v)
468. (s) 469. (n) 470. (c) 471. (e) 472. (r)
473. (h) 474. (i) 475. (f) 476. (k) 477. (w)
478. (o) 479. (q) 480. (m) 481. (t)

Mammals

482. (c) Even dogs do better than man, who can detect frequencies of up to 35,000 vibrations/second.

483. (b) Latin *mamma* means breast. All mammals have breasts.
484. (d) Hippopotamus exudes a reddish sticky sweat that turns brownish on drying and protects the skin from sun-burn. They are thus said to sweat blood.
485– These are all ungulates or hoofed animals.
489. Originally they had five toes to each foot, but during the process of evolution the number was reduced so that they could move more rapidly. The animals are:
485. Tapir — four toes on each foot but no hooves.
486. Camel — cloven feet, padded for walking on soft sand.
487. Reindeer — cloven hoof formed by nails of third and fourth toes.
488. Rhinoceros — foot reduced to three toes for more speed.
489. Zebra — middle toe nail has evolved to form the hoof.
490. (c) The term is derived from Latin and means 'of first rank'. Man is included in this order.
491. (a) From Latin *cetus* meaning large sea animal.
492. Lemur
493. Nycticebus. Also known as slow loris. The index finger is reduced to a small stub, giving the hand a wide grasp.
494. Papio. Also known as baboon. Hand is adapted for hanging from trees.
495. Pongo. Also known as orang-utan.
496. Gorilla
497. Man
498. Wolf which is carnivorous. It uses its sharp canines for stabbing and its premolars and molars for shearing.

499. Cattle (herbivorous). Canines are very much reduced, but good development of molars and premolars for chewing.
500. Man (omnivorous). Development of canines and molars balanced.
501. (a)   502. (b)
503. (d) When the father is a lion the offspring are called ligers and when the father is a tiger they are called tigons.
504. (a)   505. (c)   506. (b)   507. (d)   508. (a)
509. (c)
510. (d) Found in Madagascar, it is of the size of a cat. Its number has been reduced to about 50.
511. (c)   512. (d)   513. (i)   514. (p)   515. (m)
516. (k)   517. (g)   518. (j)   519. (a)   520. (o)
521. (l)   522. (n)   523. (b)   524. (f)   525. (h)
526. (e)   527. (c)

11. DIVERSITY OF PLANT LIFE

528. (c) About 5,000 new plant species are being discovered every year.
529. (a)   530. (c)
531. (d) Alginic acid is used in making smooth ice-creams.
532. (b)
533. (d) There are no living land plants which have descended from red or brown algae.
534. (a) It covers thousands of hectares in the Sargasso sea (in the North Atlantic Ocean), and can get entangled in the bottom of ships. It is thus a great menace to shipping.
535.   (a) and (b)
536. (a) Light reaching the greatest depth in water is in the blue-green region. Red algae contains pigments which reflect red light but completely

absorb blue-green light. Thus only red algae can be successful at ocean depths.

537. (c)   538. (b)

539. (c) The term comes from Latin *calcem* (lime) and *fuga* (flight). There plants literally "fly away" from calcium.

540. (d)   541. (d)

542. (b) From Greek *rhiza* (a root) and *omos* (alike). So literally rhizome is a "root like" structure.

543. (a) Because their undamaged underground rhizomes develop new leaves.

544. (b),(c) and (d). The grittiness of the pulp of guava and pear is due to the presence of stone cells.

545. (c) Derived from Greek *xylon* meaning wood. It forms the bulk of roots and stems of vascular plants.

546. (a) Derived from Greek *phloos* meaning bark. Phloem lies outer to xylem and forms the bark of a tree.

547. (b) From Greek *phyllon* (leaf) and *taxo* (arrange).

548. (d) This term does not imply that this wood is soft. In fact coniferous woods are quite hard.

549. (d) Botanically any ripe ovary is called a true fruit. Tomato, pea and coconut fit this description. On the other hand, in apple (also in fig), the main edible position of the fruit is the fleshy receptacle of the flower. Receptacle is the swollen tip of the stalk of the flower where all floral parts are attached.

550. (c) Derived from Greek words *peri* (around) and *karpos* (fruit). Pericarp is divided further into three layers. In mango, for instance outer skin is epicarp (or exocarp), sweet edible flesh is mesocarp and inner hard zone is endocarp.

551. (b) Derived from Greek words *parthenos* (virgin) and *karpos* (fruit). Banana is a parthenocarpic fruit.
552. (b) A composite fruit like the pineapple develops from an inflorescence by the fusion of flowers.
553. (a) In an aggregate fruit like strawberry or custard apple, each free carpel develops independently to form a bunch of fruits.
554. (d) The coco de mer, or double coconut palm, of the Seychelles may bear a seed weighing as much as 27 kg (60 lb). On the other side of the spectrum lie orchids whose seeds are only about 0.25 mm (0.01 in) long and so light that a million of them weigh only a third of a gram (0.01 oz)!
555. (b) During its growth, a delicate, net-like veil forms around the fungus — hence the name 'the lady of the white veil!'
556. Birch
557. Apple
558. Mesquite
559. Rugel's plantain
560. Sweet gum
561. Date palm
562. Mountain cedar
563. Blue succory
564. Austrian pine
565. Common wood-rush. The thing to note is that insect-pollinated plants such as apple and blue succory have sticky grains while wind-pollinated plants such as pine and birch have smooth surfaces.
566. Acuminate (when leaf tapers to a point).
567. Acute (tapers to an angle which is acute).
568. Bipinnate (a pinnate leaf with pinnate leaflets).
569. Biternate.

570. Cordate (heart-shaped).
571. Crenate (having a scalloped edge).
572. Dentate (having tooth-like projections).
573. Digitate (having divisions like fingers).
574. Elliptic
575. Falcate (like sickle).
576. Incised (having deep indentations or notches).
577. Lanceolate (shaped like the head of a lance).
578. Linear (like line).
579. Oblanceolate (shaped like a lance head but with the tapering end at the base).
580. Oblong (deviating from a circular form by being elongated in one direction).
581. Obovate (inversely ovate).
582. Obtuse
583. Orbicular (nearly circular).
584. Ovate (egg-shaped).
585. Palmate (shaped like a hand with fingers spread out).
586. Pedate (divided in a palmate manner with lobes divided into smaller segments).
587. Peltate (having the leaf stalk attached to the lower surface of the blade, instead of at the end).
588. Pinnate (having a series of leaflets on each side).
589. Pinnatifid (divided in a pinnate manner, with the divisions extending only halfway down).
590. Plicate (folded along its ribs like a closed fan).
591. Sagittate (shaped like an arrowhead).
592. Serrated (notched like the edge of a saw).
593. Spatulate (like a spoon).
594. Trifoliate (consisting of three leaflets).
595. Tripartite (divided into three parts, nearly to the base).

596. Truncate (Blunt end as if cut off). It would be interesting to collect as many leaves as possible and classify them in the above manner.
597. Alternate (placed singly at different heights along the sides of a stem).
598. Distichous (arranged alternately in two vertical rows).
599. Opposite (placed in pairs on the stem).
600. Perfoliate (stem piercing the centre of leaves).
601. Simple raceme
602. Spike
603. Panicle
604. Corymb
605. Capitulum
606. Simple Umbel
607. Compound Umbel
608. Simple monochasium
609. Compound monochasium
610. Simple dichasium
611. Compound dichasium
612. (b) The radicle gives rise to the root.
613. (a) Such fibrous systems are found in many herbaceous perennials especially the grasses.
614. (c) Many vegetable crops have tap roots, e.g., carrot, sugar-beet.
615. (d) Fusiform is a swollen root tapering at both ends.
616. (a) Napiform is a globular root tapering abruptly.
617. (b) Having a conical shape
618. (e) No definite shape
619. (c) These are storage roots occurring in clusters.
620. (b)
621. (b) In contrast seeds carry their own food reserve. Thus the seed-bearing plants have the great advantage that they can germinate in adverse surroundings.

622. (c) In lower plants food and water must be transported from cell to cell. That's why mosses seldom grow tall. The tallest species found in New Zealand measures only 60 cm.
623. (c)   624. (d)
625. (a) The blue-black 'berries' of *Juniperus communis* (The common juniper) are used for flavouring gin.
626. (b)   627. (c)
628. (a) From Greek *stelidion* (a small pillar).
629. (c)
630. (a) These cells are found mostly in leaves.
631. (c) This tissue enables adequate gaseous exchange even below water.
632. (d) Collenchyma acts mainly as a supporting tissue, especially in young stems and leaves.
633. (b) Tracheids are long, slender and tapered cells with heavily lignified walls, surrounding an empty lumen, the protoplasm having died.
634. (a) and (b) Sometimes the vascular bundles in monocotyledons may be arranged in two or more rings, but never in a single ring.
635. (b)
636.   Open collateral (as in helianthus).
637.   Closed collateral (as in zea mays).
638.   Open bicollateral (as in cucurbita).
639.   Concentric amphicribral (as in fern).
640.   Concentric amphivasal (as in dracena).
641.   Radial (as in angiosperm roots).
642. (e) Beaded roots are those which have swollen regions at frequent intervals.
643. (a) Prop roots are massive pillar-like structures.
644. (d) These are a cluster of roots which grow downward from the base of plants.
645. (b) Assimilatory roots develop chlorophyll and even photosynthesise.

646. (c) Parasitic plants develop this kind of roots whose main function is to penetrate host tissue and obtain nutrition.
647. (b)   648. (d)
649. (c) The word comes from Latin *totus* meaning whole.
650. (b)
651. (a) They are often known as air plants because they are not attached to the ground. They obtain water and mineral from the rain and from debris that falls on the support.
652. (b) Normally 4 zoospores are formed by 2 mitotic divisions.
653. (b) Asparagus, salt wort and mangrove are halophytes.
654. (d) Idioblasts may contain a variety of materials, e.g., tannins, oils, crystals, etc.
655. (a) Anthers opening towards the outside of the flower are called extrorse.
656. (b) Exine is the outer cuticularized layer. The pollen tube, which emerges during germination of the pollen grain, is an outgrowth of the intine.
657. (a) It is seen in certain algae such as the sea lettuce.
658. (c) The term is derived from French *etioler* (to blanch).
659. (b)   660. (c)   661. (b)   662. (d)
663. (a) It is a thread-like structure. The name comes from Latin *proto* (first) and Greek *nema* (thread). So literally the term means first thread.
664. (c)
665. (a) From Greek *arche* (beginning and *gonos* (race).
666. (b)
667. (d) The singular is sorus. The word comes from Greek *soros* (heap).

668. (c)   669. (d)
670. (a) The species is *pinus gerardiana*.
671. (b) The smaller spores develop into male gametophytes and the bigger into female gametophytes.
672. (d) Lignin is stained red by this.
673. (a)
674. (b) Carob tree is native to southern Europe and the Middle East.
675. (c) From Greek *leukos* (white).
676. (e) Also known as heliotropism. Shoots show this types of tropism.
677. (c) Roots show this type.
678. (d) Hyphae of certain fungi such as *mucor* are positively chemotropic, i.e., they grow towards a particular source of food.
679. (b) It is nothing but a special type of chemotropism. Roots are positively hydrotropic, the stimulus of water being stronger than the stimulus of gravity.
680. (a) Also known as haptotropism. Tendrils of climbing plants are thigmotropic. The word comes from Greek *thigmatos* (touch).

Fungi

681. (a) From Greek *sapros* (rotten).
682. (b) Also known as potato blight. It destroyed potato on a mass scale on which the Irish population was mainly dependent. Many people starved to death and many others emigrated to the USA.
683. (c) This mould is also known as *Rhizopus*.
684. (a) Name derived from Greek *askos* meaning skin bag. The distinctive feature of this phylum is a sac-like structure in which ascospores develop.
685. (d) Basidiomycetes are the most commonly seen fungi.

686. (a) It belongs to ascomycetes.
687. (c)   688. (b)   689. (b)
690. (c) Chitin is a nitrogen-containing polysaccharide, present also in cuticles of insects.
691. (a) Consequently all mitotic events take place within the nuclear envelope.
692. (d) This is the baker's yeast.
693. (b)
694. (a) Newspaper is made from barks of trees, so it is mainly cellulose.
695. (a) It gives the familiar penicillin.
696. (c) From Greek words meaning "fungus root". Non-mycorrhizal plant is an exception rather than a rule. 90% of all higher plants are mycorrhizal.
697. (d)
698. (d) They look rather like gills of fishes.
699. (a)
700. (d) Although it is closely related to *penicillium glaucum*, *penicillium notatum* has never been observed to reproduce sexually.
701. (d) Clement VII was Pope of the Roman church from 1523 to 1534.
702. (b) It can live in jet fuel tanks and use the fuel as a source of food.
703. (a) Black périgord truffle sells for as much as Rs.26,000 per kg. Still rarer white variety sells for as much as Rs.70,000 per kg.
704. (b) Dogs come a close second.
705. (b)

12. GREAT EXPERIMENTS

706. (b) Redi (1626-1697) was an Italian biologist. Incidentally, he was a poet too!

707. (d) This is achieved by juggling with genes. They now hope to breed a cow weighing 4.5 tonnes — almost as big as the elephant.
708. (a) Harvey (1578-1657) was an English physician.
709. (c) Schwann (1810-1882) was a German biologist.
710. (b) The meat was digested by stomach juices entering through wire gauze coverings.
711. (d)
712. (a) This experiment was a fatal blow to Lamarckianism because according to Lamarckian interpretation of this experiment, penicillin somehow induced a change in the bacteria enabling them to grow in the presence of penicillin. If this were true, all bacteria should be able to thrive in penicillin.
713. (b) His work was recognized 27 years later when he received Nobel Prize in Medicine in 1956.
714. (c) Forssman was barred from hospital appointments and he had to resort to private practice.
715. This represents Edward Jenner (1749-1823) vaccinating his own baby son against smallpox. This daring experiment ultimately led to the eradication of smallpox from the whole world.
716. (a) This observation established the existence of the testicle's internal secretion and was a landmark in endocrinology.
717. (b)

13. GREAT BIOLOGISTS

718. (a) Goethe (1749-1832) is more famous as a poet, but he was an accomplished scientist too.
719. (a)   720. (c)   721. (d)
722. (a) This was around 500 B.C.
723. (b)
724. (d) Fernel (1497-1558) also published a book on the subject in 1542.

725. (b) Hansen did it in 1874.
726. (a) Coincidentally Fabricius (1537-1619) was Harvey's teacher.
727. (b) Harvey measured blood flow mathematically.
728. (c) Malpighi (1628-1694) was an Italian physiologist.
729. (b) This was in 1661.
730. (a)
731. (c) Swammerdam was just 21 years of age at that time.
732. (b) They are now also known as Graafian follicles.
733. (d) Brown (1773-1858) was a Scottish botanist.
734. (a) This was in 1831.
735. (b)
736. (c) Golgi (1843-1926), an Italian biologist discovered this apparatus, which is now known as the Golgi apparatus.
737. (a)
738. (d) Oparin (1894-1980), the Russian biochemist, ultimately led others to accept that the problem of the origin of life is susceptible to intellectual and experimental investigation.
739. (a) Monod's major contributions lay in elucidating the mechanisms of control of gene expression. For this work he was awarded the Nobel Prize in 1965.
740. (c) Cuvier also had the remarkable ability of reconstructing a whole animal from only a few available bones!
741. (b) He worked out the neural structure of brain and spinal cord and also of the retina. Cajal (1852-1934) was a Spanish biologist.
742. (d)
743. (a) Wolff was the first to show that specialised organs arose out of unspecialised tissue.

744. (b) Gesner (1516-1565) is also regarded as the founder of modern zoology.
745. (a) Despite his irrational views, Oken (1779-1851) helped science in a way. His ideas included the idea of evolution well before Darwin's time.
746. (c) Driesch (1867-1941), a German embryologist, towards the end of his life became interested in occult and became a prominent member of thc British Society for Parapsychology.
747. (b) The drug synthesized in 1909 was called salvarsan (safe arsenic).
748. (a) Theophrastus (372 B.C.-287 B.C.) described over five hundred species of plants.
749. (b)
750. (c)
751. (a) Semmelweiss (1818-1865) made great enemies as a result of this, despite dramatic lowering of the incidence of childbed fever. He was once even kept in a lunatic asylum.
752. (d) Jenner, an English doctor, had discovered vaccination in 1796. Napoleon's honour is striking because England and France were at war during that time.
753. (b) Bichat (1771-1802), the French physician, was the first to use the ward tissue in a biological sense and identified 21 types of tissues in all.
754. (c) His discovery of chicken cholera vaccine was made by chance when he accidentally left some bacterial cultures unattended for several months.
755. (a) He did this macabre experiment in 1901. The fractures of facial bones are now known as Le Fort fractures.
756. (c) Banks (1743-1820) made this voyage in 1768 in the ship *Endeavour* whose commander was the famous English navigator James Cook.

757. He is Edward Jenner (1749-1823), the English physician who perfected the vaccination technique in 1796.

758. He is Lord Joseph Lister (1827-1912), the English surgeon who is considered the father of antiseptic surgery.

759. He is Louis Pasteur (1822-1895), the French scientist considered the founder of bacteriology.

760. He is Sir Ronald Ross (1857-1932), the British physician who was awarded the Nobel Prize in 1902 for studies in malaria. He was born in Almora, India.

761. He is Sir Alexander Fleming (1881-1955), the British bacteriologist who discovered penicillin in 1928.

762. He is Wilhelm Röntgen (1845-1923), the German physicist who discovered X-rays.

763. He is Paul Ehrlich (1854-1915), the German bacteriologist who synthesized the drug salvarsan for syphilis.

764. (b) Ingenhousz (1730-1799) was a Dutch physician and plant physiologist.

765. (a) Calvin described a series of chemical reactions taking place during photosynthesis. This series is now known as Calvin cycle.

766. (d)   767. (b)

14. BIOLOGY IN EVERYDAY LIFE

Agriculture

768. (b)   769. (c)   770. (d)   771. (b)

772. (a) They contain several atoms of chlorine per molecule

773. (d)   774. (c)   775. (b)

776. (a) The part that becomes the supporting portion (usually the root) is called the stock. This

method combines the good qualities of both the plants.
777. (d) Mastic is also used by actors for attaching false beards, and was used in ancient Egypt for embalming.
778. (d) This is because maize has evolved under domestication to the point where it is completely dependent upon humans for survival.
779. (c) This is done to avoid self-pollination.
780. (a)

Animals in the service of man
781. (d) There is evidence that the Stone Age man had hunting dogs as far back as 10,000 B.C. and Egyptians definitely used hunting dogs 8,000 years ago.
782. (a) The last wild auroch was seen in Poland in 1627.
783. (b) From Latin *sericum* (silk).
784. (c) Used for producing honey and wax.
785. (d) From Latin *piscis* (fish).
786. (c) From Latin *aqua* (water).
787. (b) Dates for other animals are: Goat 8000 B.C., Cattle 5500 B.C., Sheep 5000 B.C., Cat and chicken 2000 B.C., Duck and Goose, 1500 B.C. and Rabbit A.D. 1000.
788. (c)

Biotechnology
789. (b)
790. (c) This technology utilizes chemical reactions naturally occurring in living organisms.
791. (a) It is produced by fermentation done by yeast.
792. (a) and (b). Stomach wall contains an enzyme 'rennet' and fig sap contains 'ficin'. Both enzymes curdle the milk and produce cheese.

793. This is the biochip, which consists of a semi-conducting organic molecule inserted into a protein framework. The whole unit is then fixed on a protein support. Biochips can be made much smaller than traditional silicon chips. They will find use in medicine, such as for regulating heart-beat.

794. (d)

795. (b) They were awarded the 1984 Nobel Prize in Medicine for developing the technique to produce hybridomas.

Medical applications

796. (a) Miss Denise Ann Darvall was 25 years old when she died. Louis Washkansky, a 53-year-old man, was the recipient of her heart and he lived for 18 days after the operation. Thus it was the first ever case in history of a man literally winning the heart of a woman!

797. (c) Scales was a British bio-mechanical engineer.

798. (d) Developed in February 1979 by a Japanese doctor, Ryochi Naito, it will help in preventing the spread of blood-transmitted diseases such as AIDS.

799. All. Jarvik 7 and Jarvik 8 were developed in 1976 and 1986 respectively by Dr. Robert Jarvik. Pen State was developed by Dr. William Pierce, and Buecherl System by Prof. Emil S. Buecherl of Berlin.

800. (b) On 2 December 1982, Jarvik 7 was implanted in him.

801. (d) Also known as NMR, it utilizes the fact that each proton in a patient's body acts as a tiny magnet.

802. (a) Also known as CAT scan, it was developed in 1972.

803. This is balloon angioplasty or percutaneous transluminal coronary angioplasty (PTCA). In this technique, a blocked artery of the heart, the coronary artery, is opened again by passing a balloon within it and inflating it. The balloon can be inserted through the arm or groin vein and thus operation is avoided.

804. (a) Enders got the Nobel Prize in 1954 for being able to grow poliomyelitis virus artificially, which had been a major bottleneck towards development of polio vaccine.

805. These are the bands produced as a result of DNA fingerprinting. It is a recent technique in which DNA of several people can be matched. The technique is used in detection of crime and establishment of paternity.

806. This is the first ever X-ray taken. It was taken by Röntgen himself of his wife's hand.

807. (c)

808. (a) Doctor Senning of Sweden invented it in 1958

809. (b) Drinken, an American professor at Harvard University, designed the iron lung in 1927.

810. (d) Einthoven (1860-1927) received the 1924 Nobel Prize for this invention.

811. (c) This technique is now known as electroencephalography (EEG). Berger (1873-1941) used his young son as a subject.

Bioenergy

812. (b)   813. (d)   814. (a)

815. (b) HMP is used in such things as washing clothes, or working in the fields.

816. (d) India has about 84 million work animals. If each animal generates 0.5 horsepower (hp), as in pulling carts, then the 'installed capacity' of animals is 42 million hp or 30,500 MW!

817. (c) It is known as hardwood. Softwood (Gymnospermous) burns rapidly and for a very short period.

15. GREAT BOOKS

818. (b) It was the first well-illustrated, printed book on human anatomy and dispelled many old beliefs.
819. (c) The year 1543 is usually considered as marking the beginning of scientific revolution. In the same year Copernicus published his epoch-making book on astronomy.
820. (d) Calcar was a pupil of Titian.
821. (a)
822. (c) It was a small badly printed book, but has remained a classic of science ever since.
823. (b)   824. (b)
825. (d) It was published in the Soviet Union in 1924 and an English translation appeared in 1936. It has been said that such a book could have appeared only in the Soviet Union, because communists are not inhibited by religious scruples.
826. (a) This work was written between 1815 and 1822. It was also Lamarck who popularized the word 'biology'.
827. (c) In Haldane's time it was widely believed that Mendelism was not compatible with Darwinian evolution. Haldane's book published in 1932 showed that this was not so.
828. (a) The look appeared in three volumes between 1830 and 1833. Actually James Hutton, a Scottish geologist, had advanced the same views in 1785, but he was not taken seriously.
829. (b)

16. LIVING BEINGS AND THEIR ENVIRONMENT

830. (c) Biome is the biggest unit used by ecologists. Some of the main biomes are tundras, grasslands and deserts.
831. (a) Ecosystems vary in size from the whole world to a tiny grain of soil.
832. (b)

17. CURIOUS FACTS

833. (a) In 1827, while looking at pollen in water under the microscope, he noted that pollen moved about irregularly. This gave rise to the concept of 'Brownian motion'.
834. (b)
835. (d) Beaumont would tie a piece of meat with thread and introduce it into the wound and then take away the piece after an hour, thus directly visualizing the process of digestion.
836. (c) Sometime later the vineyard was struck by a fungal disease and all but the sprayed plants perished. Thus began the use of Bordeaux mixture (40g copper sulphate + 40g calcium hydroxide in 5 litres of water) as a fungicide. This mixture actually boosted grape production in France.
837. (a) The seeds of tambalacoque tree could only germinate, when they had first been eaten by a dodo. This was found as late as 1970s by an American ecologist, Stanley Temple. By that time only 13 tambalacoque trees were left in the world.
838. (c)
839. (c) During 1634 and 1637, the "tulipomania" was so much that one single tulip bulb could be traded for as much as 4,600 florins (about

$1800 today). This mania for tulips suddenly waned in 1637. This waning of interest in tulips has been compared with the 1929 U.S. stock market crash.

840. (b) Water lily's leaves are up to 2 m (7 ft) across, and the arrangement of its ribs gives them such strength that they can support a child. Sir Joseph Paxton, the designer of Crystal Palace, studied the leaves before building his masterpiece.

841. (a) On the contrary, the white rose of York was a form of dog rose.

842. (d) Also known as Judas tree, it is a small tree belonging to the pea family.

843. (c) Baobab *(Adansonia digitata)* grows in arid regions of central and southern Africa and produces large gourd-shaped fruits containing an edible pulp known as monkey bread.

844. (b) Also known as monkey puzzle. This tree *(Araucaria araucana)* grows wild in the Andes region of South America.

845. (a) In the last stages of ripening, the spore capsules shrink and shape like tiny gun barrels, each with its own airtight cap. Eventually the cap breaks away and the trapped air escapes with an audible 'pop', firing the packet of spores inside as much as 2m (7 ft).

846. (d) Rattan palm native to Malaysia and Indonesia winds its way snake-like through trees and can reach a length of 169 m (555 ft).

847. (b) Millions of *pyrodinium*, a type of plankton, make this bay shine in the night.

848. (a) When the engineer Georges de Mestral was returning from a day's hunting, burdock seeds clung to his clothing. Under the microscope he found that seeds contained numerous minute hooks.

849. (c) Lamarck (1744-1829) later produced his curious theories of evolution.
850. (b) Joseph Stalin forced all Soviet biologists to accept Lysenko's views. Lysenko did more damage to Soviet biology than anyone else.
851. (a) This was because he refused to accept the controversial theories of Lysenko. The trial lasted only five minutes. Vavilov died in a labour camp in 1943.
852. (c) In the summer of 1928 his bacterial cultures were accidentally contaminated by a fungus from which he extracted penicillin.
853. (d) Galvani (1737-1798) believed in 'animal electricity' but he was proved wrong by Volta, the Italian physicist.
854. (b) Also known as bog moss, it was so used because of its water-absorbent qualities.
855. (c)
856. (c) Owen (1804-1892), the English zoologist discovered them in 1852.
857. (b) This trial was widely publicized as the Scopes' trial. Although Scopes was defended by the leading lawyer of his time, Clarence Darrow, he was convicted. Some people called it the Monkey Trial.
858. (c) Analysis of tissues of such a mouse reveals random mixtures of the two original genotypes.
859. (a) He crushed yeast cells with sand to get a hypothetical enzyme 'zymase' which could ferment sucrose to ethanol.
860. (d) Compass plant is the name given to several plants which behave in this fashion. They do so in the scorching midday sun.
861. (a)

862. (d) Carob is an everygreen tree native to southern Europe and the Middle East. The weight of one seed is known as a carat.

863. (c) Romans called Hera by the name Juno. White lily was known to them as Juno's roses. Most of Juno's milk remained in the sky giving rise to the Milky Way.

864. (b) Schleiden (1804-1881) was a German botanist, who later elaborated the cell theory along with Schwann.

865. (a) Dalton (1766-1844) described this disease in 1794. The disease is often known as Daltonism.

18. THE HUMAN BODY

866. (a)   867. (c)

868. (a) It is the fan-shaped part of the hip bones in man. The spelling for the longest part of small intestine is ileum so (b) is not the correct answer!

869. (d)   870. (a)

871. (d) Named in honour of Italian anatomist Bartolomeo Eustachio (1524-74), who first described these tubes.

872. (b)   873. (b)   874. (a)

875. (c) Water makes up some 70% of the body.

876. (b)   877. (a)   878. (c)   879. (d)   880. (c)

881. (d) Saliva also contains this enzyme which attacks bacteria present in food.

882. (a) Raised body temperature inhibits the growth of micro-organisms. Thus fever during infections is helpful.

883. (c)

884. (b) B-cells produce antibodies. This is called humoral immunity.

885. (b) They are so called because they possess ducts.

886. (c) They do not possess ducts and pour their secretions directly into the blood.
887. (b)
888. (d) Named after the London anatomist Clopton Havers (1650-1702) who discovered them.
889. (a)
890. (c) This value is reported in a doctor's lab report as 15g%.
891. (d)
892. (b) Sometimes known as dorsal vertebrae too. On each thoracic vertebra is attached a pair of ribs, there being 12 pairs of ribs too.
893. (c)
894. (d) This is a group of special muscle fibres found in the heart, which transmits electrical impulses. It was first described by German physician Wilhelm His Jr. (1864-1934) in 1893.
895. (e) Helps in smelling.
896. (l) Helps in seeing.
897. (k) It controls eye movements.
898. (a) The smallest cranial nerve. Helps in eye movements.
899. (f) The largest cranial nerve.
900. (j) It helps in eye movements.
901. (h) Useful in controlling facial expressions.
902. (i) Helps in hearing.
903. (b) It supplies the tongue and the throat.
904. (g) Supplies a large number of organs, such as muscles of the heart and intestines.
905. (c) Supplies muscles of arms and shoulders.
906. (d) All these nerves arise in the brain and come out of the skull through holes in its base. You may find it difficult to remember the names. Remember this mnemonic for ease: *On old Olympus' topmost top a fat eared German viewed a hop*. This sentence contains twelve words each

beginning with a letter which will give you the name of the nerve. The eighth word is "eared". The VIII nerve is vestibulocochlear. You can remember it by remembering that vestibulocochlear nerve is an "ear" nerve. All the other nerves are straightforward to remember.

19. MISCELLANEOUS

907. (c) 908. (a) 909. (d) 910. (b) 911. (a)
912. (c)
913. (d) He was the Swedish chemist who had developed the concept of ionization.
914. (b)
915. (a) This American biologist in 1969 suggested an entirely new five-kingdom classification of life. In addition to old kingdoms of plants and animals, he suggested a separate kingdom for fungi. In addition he suggested two separate kingdoms for unicellular organisms — Monera for primitive cells such as bacteria and Protista for advanced cells such as unicellular algae.
916. (c)
917. (d) Named after Sir Stamford Raffles who leased Singapore from the Sultan of Johore in 1819. It is the largest and smelliest flower in the world. The flower is 3 ft in diameter, weighs 7 kg and smells like decaying flesh.
918. (d) Also known as bay laurel or sweet bay, its leathery, dark green, oval leaves are strongly aromatic and nowadays widely used for flavouring soups, meat and other dishes.
919. (a) The tallest Wellingtonia in Britain is a 50-m (165-ft) tree at Endleigh, Devon.
920.   This is the electron microscope.

921. (d) A very apt remark indeed. The total living animal species are estimated to be 1,200,000 of which 300,000 species are those of beetles only!
922. (c) Mostly it is the bog moss.
923. (b) When Huxley read *The Origin of Species* he became at once an ardent exponent of Darwinism and is supposed to have remarked, "Now why didn't I think of that?"
924. (d) In short, Gaia hypothesis states that living organisms control earth's environment to suit them. The theory is controversial.
925. (b) Typical haplonts are filamentous green algae such as *ulothrix*.
926. (d)
927. (c) Sweat glands, tear glands and mammary glands are examples of exocrine glands.
928. (c) It is a polymer of fructose. *Dahlia* root tubers contain it.
929. (d)
930. (a) Also known as alloantigen. Isoantigens are responsible for the rejection reaction to tissues transplanted within the same species.
931. (b)   932. (c)   933. (a)
934. (b) Vitamins actually help to match an enzyme to a substrate so that the reaction can take place.
935. (b) Aestivation occurs most commonly among invertebrates, but is also seen in higher animals such as tenrec, which is a mammal.
936. (b)   937. (a)   938. (c)   939. (a) and (b)
940. (b) The embryo was formed in a test tube and preserved for two months in liquid nitrogen at a temperature of $-196°C$.
941. (d)   942. (d)   943. (a) and (b)   944. (b)
945. (c)   946. (d)   947. (b)   948. (a) and (b)
949. (c)   950. (a)   951. (d)   952. (b)   953. (a)
954. (d)   955. (a)

956. (d) Colostrum is rich in nutrients and antibodies.
957. (c)   958. (b)   959. (b)
960. (c) Hydrophobia means fear of water, A rabies patient is unable to drink water.
961. (d)
962. (a) Named after Danish physiologist Christian Bohr (1855-1911) who discovered this effect.

20. ONE-LINERS

963. (T)
964. (F) Among men the incidence of colour blindness is 8% while among women, it is only 0.5%
965. (T)
966. (T) Rabies is usually spread by bites from infected (mad) dogs but it is also spread by bites of other animals such as foxes, wolves, jackals, mongooses, cats, skunks and even bats and coyotes.
967. (T) SQUID stands for Superconducting Quantum Interference Device and MET stands for Magnetoencephalography. These techniques detect weak magnetic fields of the brain.
968. (F) Purkinje fibres are found in heart. They spread the electrical activity to the walls of ventricles.
969. (T)
970. (F) It is found in forearm.
971. (F) Radiocarbon dating can only be used for specimens less than 7,000 years old. Older specimens are dated by potassium-argon dating.
972. (T)
973. (T) Patella over the knee joint in man is the largest sesamoid bone.
974. (F) Cro-Magnon man is a relatively recent man. He lived in Europe from about 40,000 years ago till about 13,000 years ago.

975. (F) Only dead material can be studied in an electron microscope because the specimen must be in a vacuum and electrons eventually heat and destroy the material.

976. (F)

21. PHOTO QUIZ

977. This diagram appears in William Harvey's classical book, *De Motu Cordis* (1628). It demonstrates one way valves in veins. A tourniquet is tied so that veins swell up as in Fig 1. When the vein is blanched and its far end blocked with a finger, the blood from near end cannot go back to fill up the void. This demonstrates the presence of valves.

978. These are starch grains found in different plant tissues. Their shapes in each plant is characteristic. There is a centre called the hilum around which layers of starch get deposited. Sometimes there may be more than one centre of deposition. Such grains are called compound grains. Semi-compound grains are those when deposition begins about two or more centres, but finally there is a common layer. A certain number of grains in potato starch are of this type.

979. This is a section through the adrenal gland which sits over the kidneys.

980. Man
981. Roundworm
982. Lobster
983. Chick
984. Funaria
985. Newt
986. Frog
987. Snail

988. Fern
989. Earthworm
990. This series depicts the microscopic hydra catching its prey and swallowing it.

22. PICTURE QUIZ

991. This is sea anemone belonging to phylum Cnidaria. Over a thousand species are known. They are usually found attached to a rock, but some live in partnership with hermit crab, attaching themselves to the crab's shell. The crab receives protection from the stinging cells of the sea anemone, which in turn obtains food from the host.
992. These are rod-shaped bacteria sitting on the sharp edge of a pin. The edge of the pin appears like a pink mountain.
993. It is an influenza virus.
994. This is the bird of paradise hanging upside down at the climax of his courtship display.
995. This is a rare species of orchid, *rhizanthella gardneri*. It remains underground, living off the decaying stumps of the broom honey myrtle, a shrub.
996. This is the pencil sea urchin, belonging to the phylum echinodermata.
997. This is the saguaro cactus, growing to a height of 60 ft.
998. These are dandelion seeds. Numerous fire hairs over them enable them to travel large distance.
999. This is a supreme example of mimicry in the plant world. There is no insect here. This is the flower of the yellow bee orchid which resembles a female bee. This attracts males, who collect pollen while trying to mate.

1000. He is Dr Lalji Singh (b. 1947), the famous molecular biologist, who started DNA fingerprinting in India.

Printed in the USA
CPSIA information can be obtained
at www.ICGtesting.com
LVHW021259041024
792814LV00015B/761